Microhardness of polymers

This book deals with the micromechanical characterization of polymer materials. Particular attention is given to microhardness measurement as a technique capable of detecting a variety of morphological and textural changes in polymers. A comprehensive introduction to the microhardness of polymers is provided, including descriptions of the various testing methods in materials science and engineering. The book also includes the micromechanical study of glassy polymers and discusses the relevant aspects of microhardness of semicrystalline polymers. It is demonstrated that microhardness, in combination with other techniques such as microscopy, differential scanning calorimetry and X-ray scattering can be very helpful in better understanding structure–property relationships, and in doing so can contribute to the improvement of physical and mechanical properties, manufacturing processes and design of parts made from polymeric materials. This book will be of use to graduate level materials science students, as well as research workers in materials science, mechanical engineering and physics departments interested in the microindentation hardness of polymer materials.

Born in 1936, Professor BALTÁ CALLEJA was educated in Spain where he obtained his first degree in physics at the University of Madrid. In 1958 he started his research work at the University of the Sorbonne in Paris on pioneering NMR studies of organic liquids relating to intermolecular effects. In 1959 he moved to the H.H. Wills Physics Laboratory in Bristol to work on crystallization and morphology of synthetic polymers. In 1962 he obtained a PhD in physics at the University of Bristol. In 1963 he was appointed Adjoint Professor of Electricity and Magnetism at the University of Madrid. Since 1970 he has led a group on macromolecular physics at the Spanish Research Council. Presently, he is Professor of Physics and Director of the Institute for Structure of Matter, CSIC, in Madrid. He is Chairman of the Macromolecular Board of the European Physical Society and has also been chairman of the Solid State Physics Group of the Spanish Royal Society of Physics. He was awarded both the Humboldt Research Award and the DuPont Research Award in 1994. He is, or has been, a member of the editorial boards of *Acta Polymerica*, *Journal of Macromolecular Science – Physics* and *Journal of Polymer Engineering*. He is the author of more than 280 papers and has contributions in several books. He is the author of the book *X-ray Scattering of Synthetic Polymers*, Elsevier, 1989.

Born in 1936, Professor FAKIROV was educated in Bulgaria where he obtained his first degree in chemistry at the University of Sofia in 1959. In 1960 he was appointed Assistant Professor at the same university. In 1962 he moved to the Lomonosov State University, Moscow, obtaining his PhD degree in chemistry in 1966. In 1972 he was appointed Associate Professor at Sofia University. In 1987 he was appointed full Professor of Polymer Chemistry at the same university. He has spent many years as Visiting Professor in various international institutions. Since 1984 he has led a group on the structure and properties of polymers at the University of Sofia. He has been Vice President of Sofia University. He was given the Award of the Union of Bulgarian Scientists and the Honour Award '100 Years of the University of Sofia'. He is a member of the editorial boards of *Applied Composite Materials* and *Advances in Polymer Technology*, and Associate Editor of the *International Journal of Polymeric Materials*. He is presently a fellow of the Foundation for Science and Technology, Portugal, and has also been a fellow of many other international institutions. He is the author of more than 220 papers, has contributions in several books as editor and coauthor, and holds nine US patents. He is the author of the book *Structure and Properties of Polymers*, Martinus Nijhoff, Int., 1985.

Cambridge Solid State Science Series

Editors

Professor D. R. Clarke
Department of Materials, University of California, Santa Barbara

Professor S. Suresh
Department of Materials Science and Engineering,
Massachusetts Institute of Technology

Professor I. M. Ward FRS
IRC in Polymer Science and Technology, University of Leeds

Titles in print in this series

To our wives
Sabine and Galina,
for their year-long
patience, understanding and support

The authors

Microhardness of polymers

F.J. Baltá Calleja
Instituto de Estructura de la Materia, CSIC, Madrid

S. Fakirov
University of Sofia, Sofia

CAMBRIDGE UNIVERSITY PRESS
Cambridge, New York, Melbourne, Madrid, Cape Town, Singapore, São Paulo

Cambridge University Press
The Edinburgh Building, Cambridge CB2 8RU, UK

Published in the United States of America by Cambridge University Press, New York

www.cambridge.org
Information on this title: www.cambridge.org/9780521642187

First published 2000
This digitally printed version 2007

A catalogue record for this publication is available from the British Library

Library of Congress Cataloguing in Publication data

Baltá Calleja, F.J.
    Microhardness of polymers / F.J. Baltá Calleja and S. Fakirov
        p.   cm.   –   (Cambridge solid state science series)
    Includes bibliographic references and index.
    ISBN 0 521 64218 3 (hc.)
    1. Polymers – Mechanical properties. 2. Microhardness. 3. Polymers – Testing.
    I. Fakirov, Stoiko. II. Title. III. Series.
TA455.P58B25   2000
620.1′9204292–dc21   00-027754   CIP

ISBN 978-0-521-64218-7 hardback
ISBN 978-0-521-04182-9 paperback

Contents

Preface

The search for quantitative structure–property relationships for the control and prediction of the mechanical behaviour of polymers has occupied a central role in the development of polymer science and engineering. Mechanical performance factors such as creep resistance, fatigue life, toughness and the stability of properties with time, stress and temperature have become subjects of major activity. Within this context microhardness emerges as a property which is sensitive to structural changes.

The microindentation hardness technique has been used for many years for the characterization of such 'classical' materials as metals, alloys, inorganic glasses, etc. Its application to polymeric materials was developed in the 1960s. The potential of this method for structural characterization of polymers was developed and highlighted to a large extent by the studies carried out in the Instituto de Estructura de la Materia, CSIC, Madrid.

Nowadays, the microhardness technique, being an elegant, non-destructive sensitive and relatively simple method, enjoys wide application, as can be concluded from the publications on the topic that have appeared during just the last five years – they number more than 100, as is shown by a routine computer-aided literature search. In addition to some methodological contributions to the technique, the microhardness method has also been successfully used to gain a deeper understanding of the microhardness–structure correlation of polymers, copolymers, polymer blends and composites. A very attractive feature of this technique is that it can be used for the micromechanical characterization of some components, phases or morphological entities that are otherwise not accessible for direct determination of their microhardness.

Since the 1940s studies of the microhardness of metals and ceramics have furnished a well established picture of the basic mechanisms involved, as recognized by the First International Workshop on Indentation held in Cambridge in 1996. However, the investigation of the microhardness of polymers is more recent and has so far been restricted to aspects of rather technical interest. The test has been used to examine: the correlation of rheological properties with microhardness, the comparison of yield stress from hardness and tension tests, the determination of internal stresses in the surface of plastics, lacquer coatings, the curing of resins and the diffusion of water through nylons. However, the lack of fundamental knowledge of the influence of the structure (lamellar thickness, crystallinity, chain orientation, etc.) on microhardness has only been recently recognized.

The microhardness technique has been established as a means of detecting a variety of morphological and textural changes in crystalline polymers and has been extensively used in research. This is because microindentation hardness is based on plastic straining and, consequently, is directly correlated to molecular and supermolecular deformation mechanisms occurring locally at the polymer surface. These mechanisms critically depend on the specific morphology of the material. The fact that crystalline polymers are multiphase materials has prompted a new route in identifying their internal structure and relating it to the resistance against local deformation (microhardness).

After an introductory chapter highlighting the value of microhardness in science and engineering, the techniques for the determination of microhardness of polymeric materials together with data evaluation are discussed in Chapter 2. Chapter 3 is devoted to the study of glassy polymers and attention is focused on glass transition temperature determination, ageing of glassy polymers and micromechanics studies of amorphous polymers with applications to the study of crazing in polymer glasses. Chapter 4 deals with the microhardness of crystalline polymers. The effects of various crystal parameters such as the degree of crystallinity, crystal size, surface energy, etc., as well as the influence of the molecular weight are discussed. The detection of polymorphic transition is also discussed. In this chapter examples showing the possibility of following the crystallization kinetics by means of microhardness are also given. The relationship between hardness and macroscopic mechanical properties is briefly presented in the light of various mechanical models. Chapter 5 covers the microhardness of polymer blends, copolymers and composites – all of which are multicomponent and/or multiphase systems and, therefore, showing often some peculiar behaviour in comparison to the homopolymers. In Chapter 6 more recent data on microhardness under strain are summarized and the concept of reversible microhardness is introduced. Finally, in Chapter 7 examples are given of the application of the microhardness technique for the characterization of polymeric materials including the influence of processing conditions, the characterization of natural polymers and biopolymers, weathering tests, the characterization of modified polymer surfaces and others.

It is noteworthy that, in contrast to some other books in the same series of Cambridge University Press, the present one is based not on a lecture course but mainly on the systematic investigations carried out during the 1980s and 1990s at the Instituto de Estructura de la Materia, CSIC, Madrid. We wish to record our appreciation here to Drs F. Ania, M.E. Cagiao, T.A. Ezquerra, A. Flores and D.R. Rueda for their many helpful discussions, comments and suggestions and for their invaluable participation in much of the work described in this volume. The book also has another peculiarity. Although it does not pretend to review all the publications on microhardness of polymers, it is based on our own studies on polymer structure in which have been involved a great many of the leading polymer scientists world-wide, including, to name only some: Professors A. Keller, I.M. Ward and D.C. Bassett from England; H. Kilian, H.G. Zachmann, E.W. Fischer, G.H. Michler, R.K. Bayer and N. Stribeck from Germany; A. Peterlin, R.S. Porter and J.C. Seferis from the USA; T. Asano and M. Hirami from Japan; H.H. Kausch from Switzerland. Furthermore, it has been the goal of the authors to stress the correlation between the microhardness of polymers and their structure using both their own research experience and results from many other workers in the field. Nevertheless, the authors hope that the book will be not only of interest to scientists and engineers but also useful for students reading for BSc or BEng degrees in materials science and materials engineering.

We would like to thank the many friends and colleagues who have allowed us to use examples from their work and also their publishers who have given us permission to reproduce photographs and diagrams. Acknowledgement is made to various firms for the provision of illustrations of their microhardness testing devices. The majority of the work on microhardness and the structure–property relationship of polymers, performed in the IEM, CSIC, Madrid, has been supported by the Dirección General de Investigación Científica y Técnica (DGICYT) and the New Energy Industrial Development Organization (NEDO), Japan. This support has been, and still is, invaluable and is warmly acknowledged.

The authors are indebted to Professor I.M. Ward for his continual encouragement and for his helpful comments on the typescript. Thanks are also due to Mrs Ana Montero for her help during preparation of the typescript.

Last but not least, one of us (S.F.) deeply appreciates the tenure of a Sabbatical Grant from DGICYT, Spain, spent in the Instituto de Estructura de la Materia, CSIC, Madrid. He acknowledges also the support of DFG, Germany, through Grant DFG-FR 675/21-2 as well as the hospitality of the Institute of Composite Materials, Ltd, Kaiserslautern, where this project was finalized.

Madrid 1999

F.J. Baltá Calleja
S. Fakirov

Chapter 1

Introduction

1.1   Hardness in materials science and engineering

For about a century engineers and metallurgists have been measuring the hardness of metals as a means of assessing their general mechanical properties. How can one define the hardness of a material? An interesting remark in this respect was made by O'Neil (1967) in his introductory essay on the hardness of metals and alloys. He wisely pointed out that hardness, 'like the storminess of the seas, is easily appreciated but not readily measured'.

In general hardness implies resistance to local surface deformation against indentation (Tabor, 1951). If we accept the practical conclusion that a hard body is one that is unyielding to the touch, it is at once evident that steel is harder than rubber. If, however, we think of hardness as the ability of a body to resist permanent deformation, a substance such as rubber would appear to be harder than most metals. This is because the range over which rubber can deform elastically is very much larger than that of metals. Indeed with rubber-like materials the elastic properties play a very important part in the assessment of hardness. With metals, however, the position is different, for although the elastic moduli are large, the *range* over which metals deform elastically is relatively small. Consequently, when metals are deformed or indented (as when we attempt to estimate their hardness) the deformation is predominantly outside the elastic range and often involves considerable plastic or permanent deformation. For this reason, the hardness of metals is bound up primarily with their plastic properties and only to a secondary extent with their elastic properties. In some cases, however, particularly in dynamic hardness measurements, the elastic properties of the metals may be as important as their plastic properties (Tabor, 1951).

Hardness measurements usually fall into three main categories: scratch hardness, indentation hardness and rebound or dynamic hardness.

Scratch hardness

Scratch hardness is the oldest form of hardness measurement and was probably first developed by mineralogists. It depends on the ability of one solid to scratch another or to be scratched *by* another solid. The method was first put on a semiquantitative basis by Mohs (1882) who selected ten minerals as standards, beginning with talc (scratch hardness 1) and ending with diamond (scratch hardness 10).

The Mohs hardness scale has been widely used by mineralogists and lapidaries. It is not, however, well suited for metals since the intervals are not well spaced in the higher ranges of hardness and most harder metals in fact have a Mohs hardness ranging between 4 and 8.

Another type of scratch hardness which is a logical development of the Mohs scale consists of drawing a diamond stylus, under a definite load, across the surface to be examined. The hardness is determined by the width or depth of the resulting scratch; the harder the material the smaller the scratch. This method has some value as a means of measuring the variation in hardness across a grain boundary. In general, however, the scratch sclerometer is a difficult instrument to operate.

Static indentation hardness

The methods most widely used in determining the hardness of metals are the static indentation methods. These involve the formation of a permanent indentation in the surface of the material under examination, the hardness being determined by the load and the size of the indentation formed. Because of the importance of indentation methods in hardness measurements a general discussion of the deformation and indentation of plastic materials is given in Chapter 2.

In the Brinell test (Brinell, 1900; Meyer, 1908) the indenter consists of a hard steel ball, though in examining very hard metals the spherical indenter may be made of tungsten carbide or even of diamond. Another type of indenter which has been widely used is the conical or pyramidal indenter as used in the Ludwik (1908) and Vickers (see Smith & Sandland (1925)) hardness tests, respectively. These indenters are now usually made of diamond. The hardness behaviour is different from that observed with spherical indenters. Other types of indenters have, at various times, been described, but they are not in wide use and do not involve new principles.

Dynamic hardness

Another type of hardness measurement is that involving the dynamic deformation or indentation of the material specimen. In the most direct method an indenter is dropped on to the metal surface and the hardness is expressed in terms of the energy of impact and the size of the resultant indentation (Martel, 1895). In the Shore rebound scleroscope (Shore, 1918) the hardness is expressed in terms of the height of rebound of the indenter. It has been shown that in this case the dynamic hardness may be expressed quantitatively in terms of the plastic and elastic properties of the metal. Another method which is, in a sense, a dynamic test is that employed in the pendulum apparatus developed by Herbert in 1923. Here an inverted compound pendulum is supported on a hard steel ball which rests on the metal under examination. The hardness is measured by the damping produced as the pendulum swings from side to side. This method is of considerable interest, but it does not lend itself readily to theoretical treatment (Tabor, 1951).

In practice, the following test methods are in use for hardness determination.

Brinell

In this test a steel ball is forced against the flat surface of the specimen. The standard method (ASTM, 1978) uses a 10-mm ball and a force of 29.42 kN. The Brinell hardness value is equal to the applied force divided by the area of the indentation:

$$HB = \frac{2P}{\pi D^2[1 - \sqrt{(1 - d/D)^2}]} \tag{1.1}$$

in which P is the force in newtons; D is the diameter of the ball in millimetres; and d is the diameter of the impression in millimetres. A 20-power microscope with a micrometer eyepiece can measure d to 0.05 mm. The minimum radius of a curved specimen surface is $2.5D$. The results of the test on polypropylene, polyoxyethylene and nylon-6,6 have been interpreted in terms of stress–strain behaviour (Baer *et al.*, 1961).

Vickers

This test uses a square pyramid of diamond in which the included angles α between non-adjacent faces of the pyramid are 136°. The hardness is given by

$$HV = \frac{2P \sin(\alpha/2)}{d^2} = 1.854\frac{P}{d^2} \tag{1.2}$$

where P is the force in newtons and d is the mean diagonal length of the impression in millimetres. The value of HV is expressed in megapascals. The force is usually applied at a controlled rate, held for 6–30 s, and then removed. The length of the impression is measured to 1 µm with a microscope equipped with a filar eyepiece (Müller, 1965). Cylindrical surfaces require corrections of up to 15% (ASTM, 1978).

Knoop

Another commonly used hardness test uses a rhombic-based pyramidal diamond with included angles of 174° and 130° between opposite edges. The hardness is given by

$$HK = C\frac{P}{d^2} \tag{1.3}$$

where P is the force in newtons, d is the principal diagonal length of indentation in millimetres and C is equal to 14.23 (ASTM, 1978). The Vickers test gives a smaller indentation than the Knoop test for a given force. The latter is very sensitive to material anisotropy because of the twofold symmetry of the indentation (Baltá Calleja & Bassett, 1977).

Rockwell
In this test the depth of indentation is read from a dial (ASTM, 1978); no microscope is required. In the most frequently used procedure, the Rockwell hardness does not measure total indentation but only the non-recoverable indentation after a heavy load is applied for 15 s and reduced to a minor load of 98 N for 15 s. Rockwell hardness data for a variety of polymers are reported by Maxwell (1955) and Nielsen (1963).

Scleroscopy
In this test the rebound of a diamond-tipped weight dropped from a fixed height is measured (Maxwell, 1955; ASTM, 1978). Model C (HSc) uses a small hammer (*ca* 2.3 g) and a fall of about 251 mm; model D (HSd) uses a hammer of about *ca* 36 g and a fall of about 18 mm.

Scratch hardness
This test measures resistance to scratching by a standardized tool (ASTM, 1978). A corner of a diamond cube is drawn across the sample surface under a force of 29.4 mN applied to the body diagonal of the cube, creating a V-shaped groove; its width Λ, in micrometres, is measured microscopically. The hardness is given by:

$$HS = 10\,000/\Lambda \qquad\qquad (1.4)$$

The constant 10 000 is arbitrary.

Applicability of the tests
The Vickers and Brinell hardness scales are almost identical up to a Brinell hardness of about 5 GPa. This range covers all polymeric materials. The Brinell test is preferred for measuring the macrohardness of large pieces in which a large indentation (2.5–6 mm diameter) is acceptable. The Vickers macrohardness test ($P > 30$ N) is used mainly where the indentation is limited in size. The Vickers microhardness test ($P < 1$ N) is used mainly with small and inhomogeneous specimens. Forces down to 10 mN are suitable for most commercial instruments. There are testers that can operate down to 10 μN in conjunction with a scanning electron microscope (Bangert *et al.*, 1983). The Knoop microhardness test is more rapid than the Vickers test and more sensitive to material anisotropy. The Rockwell instrument is used in production and quality control where absolute hardness is unimportant. The scleroscope is used for specimens that cannot be removed or cannot tolerate large indentations. Values given by the scleroscope (dynamic hardness) and the

static-ball indenter correspond directly. The scratch test is used where indentation microhardness tests cannot be made close enough to determine local variations. Typical microhardness values for polymers are summarized by Boor (1947).

1.2   Microhardness in polymer science

1.2.1   Microhardness and deformation modes in polymers

The microhardness (H) of ionic and metallic crystals and polycrystalline specimens has been extensively investigated (Kuznetsov, 1957; Brookes *et al.*, 1971, 1972; O'Neil *et al.*, 1973). In these materials microhardness is essentially determined by primary slip systems which involve dislocation movements during the indentation process. In molecular (paraffinic) crystals on the other hand, typical deformation modes preferentially include displacement of the chains by shearing and tilting and eventually twinning and phase transitions (Baltá Calleja, 1976). Crystal defects (dislocations, kinks, vacancies, and so on) facilitate such a deformation but are overwhelmingly dominated by the anisotropy of mechanical forces: namely, strong covalent bonding in the chain direction and weak van der Waals forces normal to it. Thus, in low-molecular-weight materials, as a consequence of the large anisotropy of the crystal force field crystals are relatively weak, exhibiting very poor mechanical properties. Knoop indentations on the (001) planes of solution crystallized paraffin single crystals ($n = 32$–44) are often accompanied by the development of ridges along specific crystallographic directions (Baltá Calleja, unpublished). The occurrence of these roof shaped ridges implies a change from a vertical to an oblique structure as shown by Keller (1961), thus suggesting a shearing of the molecules in the ($hk0$) slip planes as one of the possible modes of plastic deformation, frequently leading to the observed final macroscopic fracture of the crystals.

The study of microhardness in polymers, in its earliest stages, was mainly restricted to applications of technological interest (Holzmüller & Altenburg, 1961; Nielsen, 1963), such as the determination of macroscopic internal stresses in the surface of plastics (Racké & Fett, 1971). The study of microindentation offers the specific advantage of being a local deformational process restricted to depths of a few micrometres thus leaving unaltered, in contrast to bulk deformation, the rest of the test sample. Since the indentation process is primarily controlled by plastic deformation (Brookes *et al.*, 1972) the microhardness value will be intimately correlated to the specific modes of deformation operative in polymers. In these macromolecular solids, one cannot explain the observed mechanical properties on the sole basis of crystal lattice and defects. The deformation modes in the crystalline polymer are predominantly determined by the arrangement and structure of the microcrystalline domains and their connection by tie molecules. The crystals restrict the mobility of the molecules in the amorphous layers, while the latter partly transmit the required forces for the break up of crystals and additionally provide for elastic recovery when the local stress field is removed (Hosemann *et al.*, 1972; Peterlin, 1987).

1.2.2   Microhardness additivity law

It is important to note in these introductory remarks that, like many mechanical properties of solids, microhardness obeys the additivity law:

$$H = \sum_i H_i w_i \qquad (1.5)$$

where H_i and w_i are the microhardness and mass fraction, respectively, of each component and/or phase. This law can be applied to multicomponent and/or multiphase systems provided each component and/or phase is characterized by its own H. Equation (1.5) is frequently used in semicrystalline polymers for one or other purpose operating with the microhardness values of the crystalline H_c and amorphous H_a phases, respectively.

Application of the additivity law (eq. (1.5)) presumes a very important requirement – each component and/or phase of the complex system should have a T_g above room temperature, i.e. it should be a solid at room temperature and thus capable of developing an indentation after the removal of the indenter. If this is not the case, the assumption $H = 0$ for the soft component and/or phase does not seem to be the best solution, although it is frequently done.

The presence in a complex system of a very soft, liquid-like component and/or phase (not having a measurable H value at room temperature) can affect the deformation mechanism of the entire system in such a way that it does not obey the additivity law (eq. (1.5)). This situation is discussed in more detail in the subsequent chapters.

1.2.3   Tabor's relation

Another motivation for measurement of the microhardness of materials is the correlation of microhardness with other mechanical properties. For example, the microhardness value for a pyramid indenter producing plastic flow is approximately three times the yield stress, i.e. $H \sim 3Y$ (Tabor, 1951). This is the basic relation between indentation microhardness and bulk properties. It is, however, only applicable to an ideally plastic solid showing no elastic strains. The correlation between H and Y is given in Fig. 1.1 for linear polyethylene (PE) and poly(ethylene terephthalate) (PET) samples with different morphologies. The lower hardness values of 30–45 MPa obtained for melt-crystallized PE materials fall below the H/Y *ca* 3 value, which may be related to a lower stiff–compliant ratio for these lamellar structures (Baltá Calleja, 1985b). PE annealed at *ca* 130 °C

($H \sim 75$ MPa) gives an H/Y ratio closer to that predicted for an ideal plastic solid. Thus plastics showing a high stiff–compliant ratio (high-crystallinity) approach the $H/Y \sim 3$ relation, whereas those with a low stiff–compliant ratio (low crystallinity) deviate from the classical plasticity theory. In PET samples one observes a similar behaviour. The mechanical properties (H, Y) improve when passing from the amorphous to the crystalline state. Smaller values for the H/Y ratio are obtained when the strain rate of the tensile test is much larger than that used in the indentation test (Baltá Calleja *et al.*, 1995). Values smaller than $H/Y = 3$ are also found when using the yield stress in compression (Flores *et al.*, 2000). This is due to the fact that $Y_{compression}$ is larger than $Y_{tension}$. The difference has been ascribed to the effect of the hydrostatic component of compressive stress on isotropic polymers including PE (Ward, 1971) (see Chapter 4).

An experimental relationship between the microhardness and elastic modulus (E) has been found for various PE materials with different crystallinity values (Flores *et al.*, 2000). It is important to realize that microhardness – the plastic deformation of crystals at high strains – primarily depends on the average thickness and perfection of the nanocrystals, whereas in the case of the modulus, the elastic response at low strains is dictated by the cooperative effects of both microphases, the crystalline lamellae and the amorphous layer reinforced by tie molecules. The

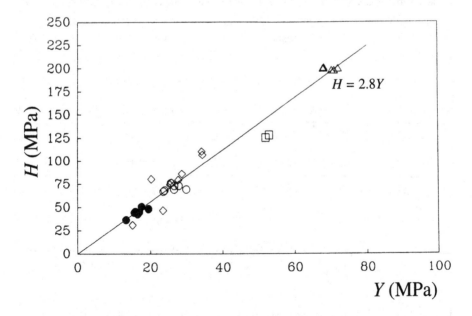

Figure 1.1. Correlation between hardness at 6 s (loading time) and yield stress of PE and PET samples with different crystallinities: (●) quenched PE at $-84\,^{\circ}$C; (○) PE slowly cooled at $4\,^{\circ}$C min^{-1}; (◊) PE annealed at atmospheric and at high pressure (4 kbar); (□) glassy PET; (△) crystallized PET. (From Santa Cruz, 1991.)

microhardness increase modulated by the chain extension of the crystals usually parallels the increase in stiffness. However, the specific dependence of H upon E for the different morphological units, gives rise to a different dependence of H upon E in systems where the rubber elastic behaviour of the amorphous layers is more pronounced (Baltá Calleja & Kilian, 1985).

1.2.4   Microhardness of polymers in contrast to metals

The common belief that a crystalline solid is always harder than an amorphous one, regardless of the chemical composition, seems to be misleading. This has been demonstrated on gelatin films (Fakirov et al., 1999). This commodity polymer, known as a very soft product in the gel state, turns out to have a very high hardness value even at elevated temperatures (150–200 °C) provided it is measured in the dry state. Its microhardness of 380–400 MPa (at room temperature with 10–15% water content, H is around 200 MPa) surpasses that of all commonly used commercial synthetic polymers and some soft metals and alloys, as can be concluded from Fig. 1.2.

   Paraffins, PE and metals, such as Pb and Sn, have microhardness values below 100 MPa. Semicrystalline polyoxymethylene, PET, chain-extended PE, poly(ethylene 2,6 naphthalate) and metals, such as Al, Au, Ag, Cu and Pt, have values between 100 and 300 MPa. The microhardness values of carbon-fibre-reinforced polymer composites are about 900 MPa and those for the common metals Zn and Co are 2000 and 4000 MPa, respectively, while for white steel it is 5000 MPa.

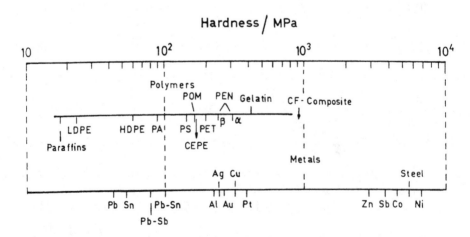

Figure 1.2. Typical microhardness values of polymers compared with data for metals. LDPE, low-density polyethylene; HDPE, high-density polyethylene; PA, polyamides; POM, polyoxymethylene; CEPE, chain-extended polyethylene; CF-composite, carbon-fibre composite; PS, polystyrene; PEN, poly(ethylene naphthalene-2,6-dicarboxylate). (From Baltá Calleja & Fakirov, 1997.)

The microhardness of thermally untreated gelatin of 400 MPa surpasses that of all commonly used synthetic polymers and soft metals and the value for the thermally treated gelatin of almost 700 MPa (Vassileva *et al.*, 1998) approaches that of the carbon-fibre-reinforced composites.

The fact that gelatin is distinguished by such a high microhardness value when in the amorphous state has another important advantage. It is known that amorphous solids are structurally ideal, i.e. they are free from structural defects unlike crystalline solids, and for this reason they have superior barrier properties.

In conclusion, let us emphasize some areas of polymer research that offer new possibilities for applications of the microindentation method to measurements of the mechanical properties of polymer surfaces. These include further microhardness–morphology correlations of flexible and rigid crystallizable polymers, microfibrillar materials and non-crystallizable glasses. Researchers interested in surface properties will recognize future opportunities in the characterization of ion-implanted polymer surfaces, coatings and weathering characterization of polymeric materials (see Chapter 7). Of particular interest is the applicability of the technique to new high-tech materials characterized by extremely high surface microhardness. Finally, it is expected that nanoindentation techniques will offer novel possibilities for studying the elastic and plastic properties of the near-surface region of polymers.

1.3   References

Annual Book of ASTM Standard Part 10, American Society for Testing and Materials, Philadelphia, 1978.

Baer, E., Maier, R.E. & Peterson, R.N. (1961) *SPE. J.* **17**, 1203.

Baltá Calleja, F.J. (1976) *Colloid & Polym. Sci.* **254**, 258.

Baltá Calleja, F.J. (1985a), *Colloid & Polym. Sci.* **263**, 297.

Baltá Calleja, F.J. (1985b), *Adv. in Polym. Sci.* **66**, 117.

Baltá Calleja, F.J. & Bassett, D.C. (1977) *J. Polym. Sci.* **58C**, 157.

Baltá Calleja, F.J. & Fakirov, S. (1997) *Trends Polym. Sci.* **5**, 246.

Baltá Calleja, F.J. & Kilian, M.G. (1985) *Colloid & Polym. Sci.* **263**, 697.

Baltá Calleja, F.J., Giri, L., Ward, I.M. & Cansfield, D.L.M. (1995) *J. Mater. Sci.* **30**, 1139.

Baltá Calleja, F.J., Martínez-Salazar, J. & Rueda, D.R. (1987) *Encyclopedia of Polymer Science and Engineering* Vol. 7 (Mark, H.F., Bikales, N.M., Overberger, C.G. & Menges, G. eds.) second edition, John Wiley & Sons, Inc., New York, p. 614.

Bangert, H., Wagendritzed, A. & Aschinger, H. (1983) *Philips Electron Optics Bull.* **119**, 17.

Boor, L., Rijan, J., Marks, M. & Bartre, W. (Mar. 1947) *ASTM Bull.* **145**, 68.

Brinell, J.A. (1900) *II. Cong. Int. Méthodes d'Essai*, Paris. For the first English account see A. Wahlberg (1901) *J. Iron & Steel Inst.* **59**, 243.

Brookes, C.A., Green, P., Harrison, P.H. & Moxley, B. (1972) *J. Phys. D. Appl. Phys.* **5**, 1284.

Brookes, C.A., O'Neill, J.B. & Redfern, B.A. (1971) *Proc. Roy. Soc. Lond. A322*, 73.

Fakirov, S., Cagiao, M.E., Baltá Calleja, F.J., Sapundjieva, D. & Vassileva, E. (1999) *Colloid Polym. Sci.* **43**, 195.

Flores, A., Baltá Calleja, F.J., Attenburrow, G.E. & Bassett, D.C. (2000) *Polymer* **41**, 5431.

Herbert, E.G. (1923) *Engineer* **135**, 390, 686.

Holzmüller, W. & Altenburg, K. (1961) *Physik der Kunststoffe*, Akademie Verlag, Berlin, p. 617.

Hosemann, R., Loboda-Cackovic, J. & Cackovic, H. (1972) *Z. Naturforsch.* **27a**, 478.

Keller, A. (1961) *Phil. Mag.* **6**, 329.

Kuznetsov, V.D. (1957) *Surface Energy of Solids*, Her Majesty's Stationery Office, London, p. 44.

Love, A.E.M. (1927) *The Mathematical Theory of Elasticity*, fourth edition, Dover Publications, London, p. 183.

Ludwik, P. (1908) *Die Kegelprobe*, J. Springer, Berlin.

Martel, P. (1895), *Commission des Méthodes d'Essai des Matériaux de Construction*, Paris, **3**, 261.

Maxwell, B. (1955) *Mod. Plast.* **32**(5), 125.

Meyer, E. (1908) *Z. d. Vereines Deutsch. Ingenieure* **52**, 645.

Mohs, F. (1822), *Grundriss der Mineralogie*, Dresden.

Müller, K. (1965) *Kunststoffe* **60**, 265.

Nielsen, I.E. (1963) *Mechanical Properties of Polymers*, Reinhold Publishing Corporation, New York, p. 220.

O'Neill, H. (1967) *Hardness Measurement of Metals and Alloys*, Chapman and Hall, London.

O'Neill, J.B.O, Redfern, B.A.W. & Brookes, C.A. (1973) *J. Mater. Sci.* **8**, 47.

Peterlin, A. (1987) *Colloid Polym. Sci.* **265**, 357.

Racké, H.H. & Fett, T. (1971) *Materialprüfung* **13**, 37.

Santa Cruz, C. (1991) PhD Thesis, Autonomous University of Madrid.

Shore, A.F. (1918) *J. Iron & Steel Inst.* **2**, 59 (Rebound sc.).

Smith, R. & Sandland, G. (1922) *Proc. Inst. Mech. Engrs.* **1**, 623; (1925) *J. Iron & Steel Inst.* **1**, 285.

Tabor, D. (1951) *The Hardness of Metals*, Oxford University Press, New York.

Vassileva, E., Baltá Calleja, F.J., Cagiao, M.E. & Fakirov, S. (1998) *Macromol. Rapid Commun.* **19**, 451.

Ward, I.M. (1971) *J. Polym. Sci.* **C32**, 195.

Chapter 2

Microhardness determination in polymeric materials

The microhardness of a polymeric material – resistance to local deformation – is a complex property related to mechanical properties such as modulus, strength, elasticity and plasticity. This relationship to mechanical properties is not usually straightforward, though there is a tendency for high modulus and strength values to correlate with higher degrees of microhardness within classes of materials. Microhardness has no simple, unambiguous definition; it can be measured and expressed only by carefully standardized tests (see Section 1.1).

Scratch tests have been used for microhardness measurements of polymeric materials (Bierbaum Scratch Hardness Test (ASTM D 1526)). These tests are related to cuts and scratches, and, to some extent, to the wear resistance of materials. Scratch tests are not always related to the resistance to local deformation and they are now being replaced by the preferred indentation test.

In the indentation test, a specified probe or indenter is pressed into the material under specified conditions, the depth of penetration being a measure of the microhardness according to the test method used. The duration of a microhardness measurement must be specified because polymeric materials differ in their susceptibility to plastic and viscoelastic deformation. An indenter will penetrate at a decreasing rate during application of the force, and also, the material will recover at a decreasing rate, reducing the depth of penetration, when the force is removed. Therefore, length of time that the force is applied must be specified. For most elastomers, the indentation will disappear when the force is removed. Consequently, the reading must be observed with the force applied. Since the measurements are dependent on the elastic modulus and viscoelastic behaviour of the material, there may be no simple conversion of the results obtained with testers of different ranges or by different methods. Also, as already mentioned, values from different indentation methods may not be related to surface microhardness resistance to scratching, or abrasion.

Hardness testing, in the past, has been mainly used as a simple, rapid, non-destructive production control test, as an indication of cure of some thermosetting materials, and as a measure of mechanical properties affected by changes in chemical composition, microstructure and ageing.

The wide range of microhardness values found in polymer materials makes it impractical to produce a single tester that would discriminate over the whole range from soft rubber to rigid plastics. For this reason, the Rockwell Tester provides many scale ranges and several types of durometers with varying forces applied to indenters of various contours (Langford, 1984).

Before starting to discuss the mechanics and geometry of indentation let us mention one of the very early publications on the microhardness of polymeric materials which used various testing techniques. Maxwell (1955) studied the indentation microhardness of plastics in an attempt to explain some of the anomalies previously noted in these measurements and to determine what physical constants of the material could be responsible for resistance to indentation. Slow-speed Rockwell-type tests were compared with high-speed rebound-type tests. Maxwell (1955) interpreted these results in terms of the rheological properties of high polymers: in particular, the elastic modulus, yield point, plastic flow, elastic recovery and delayed elastic recovery. Furthermore, he demonstrated the time and temperature dependence of the response of the material to microhardness measurements. This investigation led to the conclusion that each type of test gives some important data. However, it was also shown that the values obtained, or the relative rating of materials shown by such tests, should be used only after careful analysis of the test data from the viewpoint of the correlation of the test method with the conditions under which the materials in question should be employed.

Baer et al. (1961) later considered the indentation process in which large loads are placed on a spherical penetrator and the material beneath the indenter becomes permanently displaced. In addition, he defined the recovery process which occurs immediately after the load is released analysing it in terms of the elastic concepts developed by Hertz (Love, 1927).

Müller (1970) described the application of the microhardness technique using small loads, employing the Vickers approach. The effect of various factors on the microhardness of a wide range of polymers by means of the same approach was reported by Eyerer & Lang (1972). These authors reported that the diagonals of the impression did not change after the removal of the load.

In the last two decades the value of microhardness measurement as a technique capable of detecting a variety of morphological and textural changes in crystalline polymers has been amply emphasized leading to an extensive research programme in several laboratories. This is because microindentation hardness is based on plastic straining and, consequently, is directly correlated to molecular and supermolecular deformation mechanisms occurring locally at the polymer surface. These mechanisms critically depend on the specific morphology of the material. The fact that crystalline polymers are multiphase materials has prompted a new route in identifying their internal structure and relating it to the resistance against local deformation (microhardness).

2.1   Mechanics and geometry of indentation

Microindentation hardness is currently measured by static penetration of the specimen with a standard indenter at a known force. After loading with a sharp indenter a residual surface impression is left on the flat test specimen. An adequate measure of the material hardness may be computed by dividing the peak contact load, P, by the projected area of impression (Tabor, 1951). The microhardness, so defined, may be considered as an indicator of the irreversible deformation processes which characterize the material. The strain boundaries for plastic deformation below the indenter are critically dependent, as we shall show in the next chapter, on microstructural factors (crystal size and perfection, degree of crystallinity, etc.). Indentation during a microhardness test permanently deforms only a small volume element of the specimen ($V \sim 10^9$–10^{11} nm^3) (non-destructive test). The rest of the specimen acts as a constraint. Thus the contact stress between the indenter and the specimen is much larger than the compressive yield stress of the specimen (about a factor of 3 higher).

The indentation stresses, although highly concentrated in the plastic region immediately surrounding the contact, may extend significantly into the more remote elastic matrix. The material under the indenter thus consists of a zone of severe plastic deformation (about 4–5 times the penetration distance, h, of the indenter below the specimen surface) surrounded by a larger zone of elastic deformation. Together these zones generate stresses which support the force exerted by the indenter. The greater the penetration by an indenter the more severe is the plastic deformation in the inner zone and the larger its size.

Figure 2.1 illustrates the stress distribution on an amorphous PET sample at an indentation depth, $h = 2$ μm (Rikards *et al.*, 1998). It can be seen that the depth of the plastic zone shown is here about five times the penetration distance of the Vickers indenter.

Microhardness, therefore, appears to be an elastic–plastic rather elusive parameter (Marsh, 1964). Microhardness as a property is, in fact, a complex combination of other properties: elastic modulus, yield strength and strain hardening capacity. One way to differentiate between the reversible and irreversible components of contact deformation is to measure the elastic recovery during unloading of the indenter (Stilwell & Tabor, 1961). Extreme cases of depth recovery are best described by 'soft' metals, where it is negligible, and 'fully elastic' rubber, where it is complete.

Polymers showing a viscoelastic behaviour occupy the intermediate range. Out of all the existing hardness tests, the pyramid indenters are best suited for research on small specimens and microstructurally inhomogeneous samples (Tabor, 1951). Pyramid indenters provide, in addition, a contact pressure which is nearly independent of indent size and are less affected by elastic release than other indenters.

Figure 2.2 shows the contact geometry for a Vickers pyramid indenter with a semiangle α of 74° at zero load, at maximum load and after unloading. Several effects can be distinguished: (1) An *elastic deformation* yielding an instant elastic recovery (from B to C) on unloading. In semicrystalline polymers this effect seems to be mainly related to the elastic yielding of the amorphous component (Baltá Calleja, 1985). (2) A permanent *plastic deformation*, C, determined by the arrangement and structure of microcrystals and their connection by tie molecules and entanglements. At low strains it involves phase transformations, twin formation chain tilt and slip within crystals and at larger strains crack formation and chain

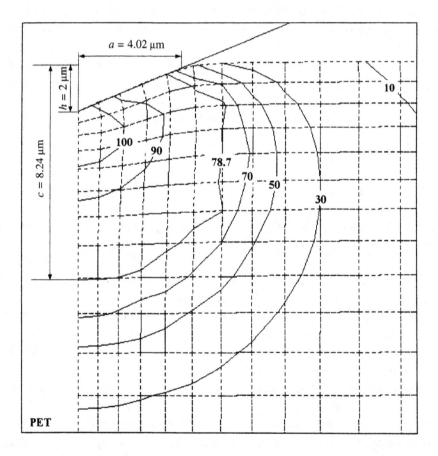

Figure 2.1. Stress distribution, in MPa, for amorphous PET at an indentation depth $h = 2\ \mu$m (c is the depth of the plastic zone and a is one-half the projected length of the indentation diagonal). Stresses larger than 78 MPa are elastic. (After Rikards *et al.*, 1998.)

unfolding (Kiho *et al.*, 1964). We may define, according to Lawn & Howes (1981), a residual impression parameter h/a. This parameter varies between 0 for an ideal elastic material ($h = 0$) and 2/7 for an ideal plastic material. (3) A time-dependent microhardness during loading (creep) (Baltá Calleja & Bassett, 1977; Baltá Calleja *et al.*, 1980b). (4) A long delayed recovery after load removal (viscoelastic relaxation) (Eyerer & Lang, 1972).

2.2   Time dependence

An important aspect concerning the surface indentation mechanism is the creep effect shown by polymeric materials, i.e. the time-dependent part of the plastic deformation of the polymer surface under the stress of the indenter (Baltá Calleja & Bassett, 1977; Baltá Calleja *et al.*, 1980b). The creep curves are characterized by a decreasing strain rate, which can be described by a time law of the form

$$H = H_0 t^{-K} \qquad\qquad\qquad (2.1)$$

The constant H_0 is a coefficient which for a given morphology depends on temperature and loading stress and K is a constant which furnishes a quantitative measure for the rate of creep of the material. A very detailed and novel description of the microscopic tensile creep in oriented PE has been given by Wilding & Ward (1978, 1981). In our investigations the H value at 0.1 min has been adopted because it approaches Tabor's relation ($H = 3Y$, see Section 1.2.3). The creep constant K for PE has been shown to depend on crystallization temperature and on annealing effects (Baltá Calleja *et al.*, 2000). This result indicates that creep depends upon, among other factors, the temperature of measurement and crystal perfection. Creep involves a viscosity element and could be associated with a crystal–crystal slip mechanism. In addition, in oriented systems, K is an anisotropic quantity $K_{\parallel} > K_{\perp}$,

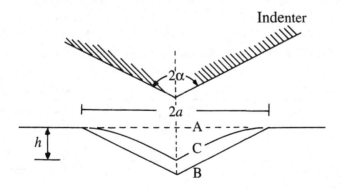

Figure 2.2. Contact geometry for a sharp indenter at: zero load (A), maximum load (B) and complete unload (C). The residual penetration depth after load removal is given by h. (From Baltá Calleja, 1985.)

independent of draw, ratio varies with the morphology of the fibrous system (Baltá Calleja *et al.*, 1980b).

2.3   Experimental details: sample preparation

As we have seen in Section 1.1, Vickers microhardness measurement uses as the indenter a square-based diamond pyramid about 100 μm in height. The included angles between opposite faces are $\alpha = 136°$. This corresponds to the tangential angle of an 'ideal' ball impression, considered to have a diameter equal to 0.375 times that of the ball (Tabor, 1951). The pyramid is pressed under a load P into the polymer surface to form a plastic deformation. When the indenter is removed the diameter of the indentation is measured and its area determined. The mean pressure p over the indentation is then

$$p = H = \frac{P}{A} \tag{2.2}$$

In the industrial test procedure A is the surface area of indentation; from the point of view of making p physically meaningful, it is more appropriate to use the projected area (see eq. (1.2)). The difference, however, is small. In the case of homogeneous surface impressions A can be measured using an image splitting eyepiece.

Specimen preparation is important in microhardness measurements and care must be taken to ensure that the microhardness recorded is representative of that material and morphology.

Consideration is given to: (*a*) possible edge effects, (*b*) the method of specimen mounting, (*c*) the problem of illumination and viewing, and (*d*) possible interactions (Bowman & Bevis, 1977).

(*a*) *Possible edge effects* The low moduli and low work-hardening coefficients of thermoplastics, in particular the polyolefins, cause any applied load to be spread over a considerably larger volume of polymer than would be the case in microhardness testing metals such as mild steel. To test the microhardness of a plastic component

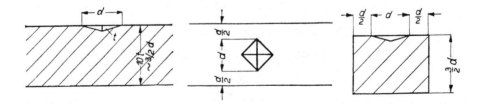

Figure 2.3. Sample size ratios for Vickers microhardness testing: the correct ratio of the diagonal length d and the thickness of a layer for a cross-section measurement has to be $2d$. (From Kleinhärteprüfer DURIMET, Ernest Leitz GmbH, Wetzlar, Liste 72-5a, 1964.)

near a free surface may result in the strain being absorbed at the specimen surface. Therefore, to ascertain the influence of the surface on the size of the indentation, the sample size ratio shown in Fig. 2.3 has to be used.

(b) Specimen mounting For successful and accurate recording of the microhardness of plastics the following conditions must be satisfied: (i) the volume of plastic receiving the indenter must be free of any external stresses (Racké & Fett, 1971) (except when the effect of the stress or strain is of interest, see Chapter 6); (ii) the surface on which the microhardness measurements are to be conducted must be parallel with the base plate of the microhardness tester; and (iii) the surface must be well polished and free of voids and scratches.

(c) Illumination and viewing The importance of obtaining a well polished surface for good viewing of the indentation has been noted by Eyerer & Lang (1972). However, over polishing to a fine mirror finish is of little use as the specimen must sit at a given distance below the indenter and this distance is set when the microscope of the indenter is focused on the test surface. The authors found that polishing with a 1 μm diamond paste polish allowed focusing. Once polished, the specimen must be illuminated and difficulty is often experienced in this respect. By depositing a thin (<50 Å) layer of Au/Pd on the plastic surface satisfactory viewing is possible. This thin layer of metal can be deposited in any conventional coating unit, and the deposited layer improves the viewing of the indentation without altering the indentation size (Bowmann & Bevis, 1977).

(d) Indentation interactions To avoid the possibility of interactions between indentations, successive indentations must be spaced well apart, in order that the plastically deformed zones around the indentation do not overlap each other nor reach an unsupported edge of the specimen. It is recommended by Bowmann & Bevis (1977) that two adjacent indentations should be not closer than the length of one diagonal of indentation.

Forces between 0.1 N and 10 N are suitable for use with most commercial microhardness instruments. However, there are also instruments which are designed to operate down to 10 μN using a scanning electron microscope (Bangert *et al.*, 1983).

As pointed out in Section 1.1 a rhombic-based pyramidal diamond (Knoop test) is also used in the microindentation study of polymers (Baltá Calleja & Bassett, 1977). Again, the microhardness value is here derived from the force applied divided by the projected area of the impression (see eq. (1.3)).

The Vickers microhardness test gives a smaller indentation for a given force than the Knoop test and is less sensitive to material anisotropy. The Knoop test is easier to use because the impression is longer for a given load and usually only one measurement per test is required. The microhardness measurement with the Knoop diamond is quite sensitive to material anisotropy because of the twofold symmetry of indentation. Nevertheless, the Vickers diamond, as we shall show below, also detects anisotropy conveniently (Baltá Calleja & Bassett, 1977; Baltá Calleja *et al.*, 1980a).

An average of several readings is normally taken. A preliminary measurement on a standard material of known microhardness is used as a control to prevent major errors.

2.3.1 Microhardness measurement on cylindrical surfaces

The test can be applied to cylindrical filaments of a very small diameter ($\cong 100$ μm) but appreciable corrections have to be made. This case was treated by Rueda *et al.* (1982) and an appropriate equation was derived in the following way.

Figure 2.4 schematically depicts the indentation geometry for a Vickers indenter penetrating a cylindrical surface with a radius r. In the case of an ideal plastic deformation (i.e. when elastic stresses are absent) after load removal, the square pyramidal indenter leaves a rhombic indentation with one of its diagonals parallel to the filament axis. Let $2BC$ be the measured indentation length, $\ell_\perp$, normal to the filament axis and $2DE$ the indentation length, $\ell_\perp^c$ which would arise on a flat surface for the same penetration depth. For an isotropic material, $\ell_\perp^c = \ell_\parallel$. However, as a result of the existing curvature, $\ell_\parallel > \ell_\perp$ (anisometric indentation). From Fig. 2.4, $\overline{DE} = \overline{BC} + \overline{BD}\tan(\alpha/2)$ and since $\tan(\alpha/2) \simeq 7/2$ one has $\ell_\perp^c = l_\perp + 7\overline{BD}$. Now, by substituting $\overline{BD} = r - \overline{BO} = r - \sqrt{(r^2 - \overline{BC}^2)}$ in the latter equation one obtains:

$$\ell_\perp^c = \ell_\perp + 7\left\{r - \sqrt{\left[r^2 - \left(\frac{\ell_\perp}{2}\right)^2\right]}\right\}$$

an expression which relates the observed diagonal $\ell_\perp$ to the ideal diagonal length $\ell_\perp^c$. The geometrical anisometry, $\Delta\ell^c = \ell_\perp^c - \ell_\perp$, calculated for a given value of

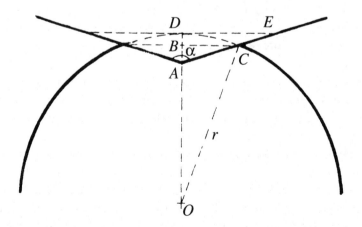

Figure 2.4. Schematic illustration of a Vickers indenter, edge on, penetrating the surface of a cylindrical material of radius, r. (After Rueda *et al.*, 1982.)

r, can now be compared with the measured indentation anisometry, $\Delta\ell = \ell_\| - \ell_\perp$. For an isotropic material, evidently $\Delta\ell^c = \Delta\ell$. If one plots $\ell^c_\perp$ against $\ell_\perp$ according to the above equation for different values of r, the bisecting straight line, $\ell^c_\perp = \ell_\perp$ corresponds to an isotropic fully flat surface ($r = \infty$) with no indentation and, for a given value of r, $\Delta\ell^c$ increases with $\ell_\perp$. However, if $\ell_\perp$ is maintained constant, $\Delta\ell^c$ decreases markedly with increasing r.

2.4 Microhardness calculation of crystalline polymers

A perfect crystalline polymer can be regarded as an efficient packing of rigid rods organized in a specific crystallographic register with the van der Waal's bonding between macromolecules weaker than intramolecular covalent linkages. The microhardness of such a system can be visualized as the resistance to local deformation produced by an external force, i.e. microhardness should be intimately related to the critical stress required to plastically deform the polymer crystal (Fig. 2.5). The applied compressive force, thus, has to separate adjacent molecules and, hence, has to overcome the cohesive forces between the said molecules. Microhardness measurements normal to the chain direction, $H_\perp$, have been performed in highly oriented samples of PE 'continuous crystals' (Baltá Calleja et al., 1980a,b). However, measurements of $H_\|$ are not frequently reported (Santa Cruz et al., 1993). If one admits that the crystal microhardness H_c in different directions follows a tensorial relationship, the average value for the isotropic polymer can be defined by:

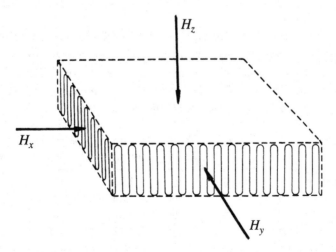

Figure 2.5. The microhardness of a polymer crystal is related to the critical stress required to overcome the cohesive forces between chain molecules. Different modes arise depending on the direction of the applied force. (From Baltá Calleja, 1985.)

$$\langle H_c \rangle = \frac{1}{3}(H_c)_\parallel + \frac{2}{3}(H_c)_\perp \qquad\qquad (2.3)$$

One may attempt to derive the ideal shear strength S_0 of the van der Waals solid normal to the chain axis from the value of the lateral surface free energy, σ. This value is well known for common polymers such as PE or polystyrene (PS) (Hoffman et al., 1976) or else can be calculated from the Thomas–Stavely (1952) relationship: $\sigma = \sqrt{a}(\Delta h_f)\gamma$, where a is the chain cross-section in the crystalline phase, Δh_f is the heat of fusion, and γ is a constant equal to 0.12. If one now assumes that a displacement between adjacent molecules by $\delta\ell$ within the crystal is sufficient for lattice destruction then the ultimate transverse stress per chain will be given by $S_0 = \sigma/\delta 1$. The values so obtained are shown in Table 2.1 for various polymers. In some cases (nylon, polyoxymethylene, polyoxyethylene (POE)) the agreement with experiment is fair. In the others, deviations are more evident. In order to understand better the discrepancy between the experimentally observed and the theoretically derived compressive strength one has to consider more thoroughly the micromorphology of polymer solids and the phenomena caused by the applied stress before lattice destruction occurs.

2.5   Crystal destruction boundary

From the foregoing is clear that the material directly under the indenter consists of a zone of severe plastic deformation. It is known that macroscopic yielding of a crystalline polymer involves a local irreversible mechanism of fracture of original lamellae into smaller units (Grubb & Keller, 1980). The heat generated during

Table 2.1. *Comparison of calculated shear strength S_0 and experimentally obtained hardness values for various polymers. The values for their respective unit cell cross-sections a and enthalpies of fusion Δh_f are also given. Lateral surface free energy $\sigma = \sqrt{a}(\Delta h_f)\gamma$. The value of S_0 depends on the choice of δl (see text). In this table $\delta l \sim 1$ Å has been chosen. (From Baltá Calleja, 1985.)*

Polymer	a (Å^2)	Δh_f (J g^{-1})	σ (erg cm^{-2})	$S_0 = \sigma/\delta l$ (MPa)	H_{exp} (MPa)
Polyethyelene	18.9	293	15.3	150	50–125
Nylon 6	17.7	230	11.6	113	90
i-Polystyrene	69.2	96	9.6	94	129-170
Polyoxymethylene	17.1	326	16.2	159	170
Poly(ethylene oxide)	16.5	196	9.5	93	70
Selenium	16.5	78.5	3.8	37	
Trans-1-4-polyisoprene	23.2	155	8.9	87	

lamellae destruction provides sufficient chain mobilization in the blocks that they rearrange to a new thickness determined by the temperature of plastic deformation (Peterlin & Baltá Calleja, 1970; Baltá Calleja *et al.*, 1972). Since microindentation also involves a yielding process we have suggested that a certain destruction and/or severe deformation of a small volume fraction of lamellae localized at the surface under the indenter may occur. This leads to a 'recrystallization' or rearrangement of the lamellae, after load removal, into a modified structure. Based on thermody-namical considerations (Baltá Calleja *et al.*, 1981) the volume of destroyed crystals under the stress field of the indenter has been approximated to:

$$V_d^c \simeq \frac{W}{\Delta \varphi} = \frac{Ph}{\Delta \varphi} \qquad (2.4)$$

where W is the mechanical work performed under the indenter, $W = Ph$ (cal) (load applied P times the indentation depth h) and $\Delta \varphi$ is the thermodynamic work per unit volume needed to destroy a stack of crystals with a given thickness ℓ. The volume, V_1, of material displaced under the Vickers pyramid, for a penetration depth, h, is given by $V_1 \sim Kh^3$. Hence, for a given load, V_d^c will be proportional to $V_1^{1/3}$. Calculations show that $V_d^c \ll V_1$ (Baltá Calleja *et al.*, 1981). Furthermore, since the microhardness of the crystals is an increasing function of their average volume it turns out that V_d^c will increase inversely with the volume of material destroyed. The linear increase of V_d^c against the reciprocal of the volume of crystallite blocks for various PE specimens is shown in Fig. 2.6. For a constant applied load P, the smaller are the crystalline blocks the larger becomes the volume displaced under the indenter and hence the larger is the crystal destruction boundary. However, from eq. (2.4) one sees that for a given penetration depth the work done will be larger the harder the material (thicker lamellae). In this case V_d^c will be larger for a softer material. Figure 2.7 shows the strain boundaries for crystal destruction round the indent (assumed to be spherical in shape) (for $h = $ constant) in the case of two PE samples with a crystallinity of $w_c = 0.95$ ($\ell_c \sim 1650$ Å) and $w_c \sim 0.21$ ($\ell_c \sim 36$ Å), respectively. In the former case (highly crystalline) the strain boundary lies at $0.18d$; in the latter case (highly amorphous material) it lies at $0.26d$.

2.6 Temperature dependence of microhardness in crystalline polymers

It may be expected that with increasing temperature the thermal expansion in the crystalline regions will lead to an enlargement of the chain cross-section in the crys-talline phase. This will induce a decrease in the cohesion energy of the crystals thus causing a gradual lowering of resistance to plastic deformation. In order to minimize the effect of the non-crystalline surface layer, the influence of the temperature on microhardness has been investigated in PE crystallized at 260 °C under a pressure of 5 kbar (Baltá Calleja, 1985). The decrease of H with temperature for the above chain-extended PE material is depicted in Fig. 2.8. The hardness decrease follows an exponential law

$$H = H_0 e^{-\beta T} \qquad\qquad (2.5)$$

where T is the temperature, H_0 is a constant and β is the coefficient of thermal softening. The exponential temperature dependence for hardness shows an inflection near $100\,^\circ$C suggesting the presence of two different mechanisms. On the lower-temperature side, changes in the thermal expansion are due to the transverse molecular motions. One may expect that deformation of chain-extended crystals

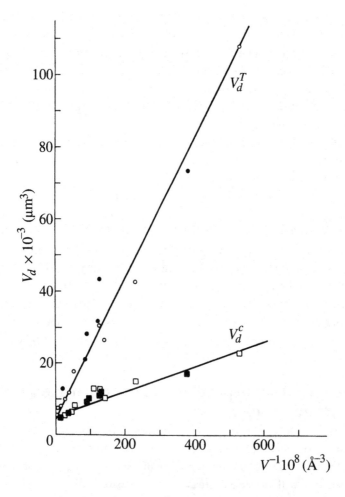

Figure 2.6. Volume of crystals destroyed, V_c^d as compared with the total volume of material plastically deformed (including non-crystalline regions) V_d^T beneath the indenter against the reciprocal of the crystal block volume. Applied load: 15 N. (After Baltá Calleja *et al.*, 1981.)

will be dominated by a chain slip mechanism. The bending of individual extended-chain lamellae seen in replicas of indented areas supports this view (Baltá Calleja & Bassett, unpublished work). For chain-folded lamellae, showing a higher creep rate, an additional interlamellar shear including chain unfolding may occur. The mechanism above the inflection point may be connected with the onset of torsional motions of macromolecular chain segments around the chain axis (Zalwert, 1970).

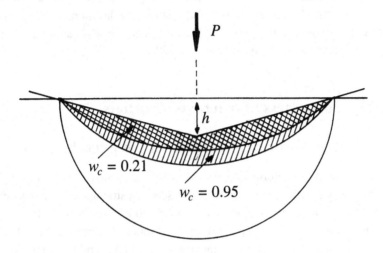

Figure 2.7. Strain boundaries for the crystal destruction zone round the Vickers indent for two PE samples with degree of crystallinity $w_c = 0.95$ and $w_c = 0.21$, respectively. (After Baltá Calleja, 1985.)

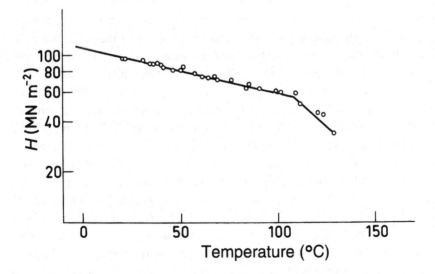

Figure 2.8. Log H measured at various temperatures for chain-extended PE ($\bar{M}_w \sim 2 \times 10^6$) crystallized at 5 kbar isothermally at 260 °C. (After Baltá Calleja, 1985.)

These motions result in partial disordering of the crystalline lattice and thus also in increased mobility of the segments at the intercrystalline regions. This leads to the change in the linear expansion coefficient.

The effect of temperature on microhardness at low temperatures (for PE between -60 and $25\,°C$ and for isotactic polypropylene (i-PP) between -20 and $25\,°C$) has also been studied (Pereña $et\ al.$, 1989; Martin $et\ al.$, 1986). While for i-PP it is possible to detect accurately the glass transition temperature, for PE it is concluded that the use of only H for recognizing the secondary relaxations in PE does not allow their precise temperature location. Here the joint use of dynamic mechanical techniques and microhardness is recommended (Pereña $et\ al.$, 1989).

2.7   Microhardness of oriented materials

2.7.1   Effect of plastic deformation on the microhardness

A uniaxial mechanical deformation provokes drastic changes in the indentation pattern of drawn polymers. Some typical results illustrating the dependence of microhardness on draw ratio for plastically deformed PE are shown in Fig. 2.9(a). These experiments (Baltá Calleja, 1976) refer to a linear PE sample ($M_w \sim 80\,000$) prepared in the usual dumb-bell form drawn at a rate of $0.5\ \mathrm{cm\ min^{-1}}$ at atmospheric pressure. The indentations were made longitudinally along the orientation axis using a Vickers pyramid. Before the neck ($\lambda = 1$), the microhardness of the microspherulitic lamellar sample is equal to ~34 MPa (loading time: 20 min); at the end of the neck ($\lambda = 5$) $H = 46$ MPa. Indentations could not be impressed in the region $1.5 > \lambda > 5$ due to the curvature of the neck. The change in H from the isotropic microspherulitic to the fibrous structure seems to be rather discontinuous, in accordance to other structural changes of polymers occurring in the neck region, such as the sudden changes in the X-ray long-period (Baltá Calleja & Peterlin, 1970; Meinel $et\ al.$, 1970), molecular orientation (Morossoff & Peterlin, 1970) and morphology (Sakaoku & Peterlin, 1971). These data are consistent with the mechanism of plastic deformation of semicrystalline polymers proposed by Peterlin (1970). A hardening of the material is thus obtained by suppressing the initial lamellar structure through mechanical deformation. The newly created fibre structure consists of highly aligned microfibrils 100–200 Å in lateral dimensions bundling into fibrils 1000–2000 Å thick. Within the microfibrils stacks of crystal blocks perpendicularly oriented to the draw direction act as cross-links for the molecules bridging adjacent crystalline layers. One may expect that microindentation hardness, which is a property related to the elastic modulus, will depend on the fraction and distribution of tie molecules in the oriented fibre structure. For PE deformed at comparatively low draw ratios ($\lambda < 8$) the tie molecules of the entangled network are relatively relaxed and most of the macroscopic indentation still occurs by deformation of amorphous regions between and within crystal blocks; similar for the mechanism occurring in the isotropic-lamellar case. For samples deformed at higher draw ratios ($\lambda > 8$) the tie molecules are strained within and between the microfibrils conferring a great stiffness and rigidity to the material. The macroscopic strain under the indenter must

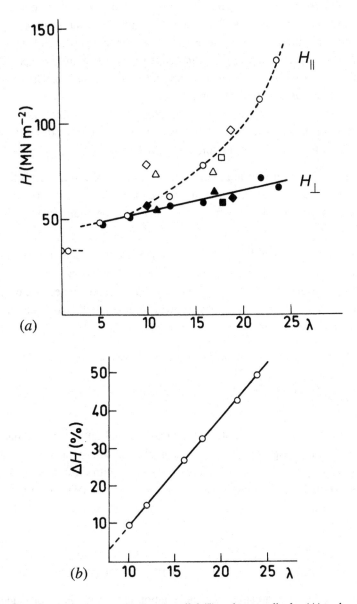

Figure 2.9. (a) Microhardness (H) values parallel ($\parallel$) and perpendicular ($\perp$) to the orientation axis as a function of draw ratio, λ, for PE drawn at different temperatures: 60 °C (O), 100 °C (−−−); 120 °C (△); 130 ° (□); (b) microindentation anisotropy ΔH of the above drawn samples vs draw ratio λ. (From Baltá Calleja, 1985.)

now occur by physical deformation of the fibrils and microfibrils. As a result, the high anisotropy of the fibre structure which increases with λ causes a conspicuous anisotropic shape of the indentation when using a square-based pyramid (Fig. 2.10). It has been shown that the anisotropy depends on the orientation of the diagonals of indentation relative to the axial direction (Baltá Calleja & Bassett, 1977). At least two well defined hardness values for high draw ratios ($\lambda > 8$) emerge. One value (the highest) can be derived from the indentation diagonal parallel to the fibre axis. Another one (the lowest) is deduced from the diagonal perpendicular to it. The former value is, in fact, not a physical measure of hardness but corresponds to an instant *elastic recovery* of the fibrous network in the draw direction. The latter value defines the *plastic component* of the oriented material.

From the morphology of the fibrous structure of the deformed polymer (see Fig. 2.11(*a*)) one may conclude that the dominant deformation modes of the drawn polymer under the stress field of the indenter involve:

(i) a sliding motion of fibrils and microfibrils, which are sheared and displaced normally to the fibre axis under compressive load. This sliding motion of microfibrils is opposed by the friction resistance in the boundary (Fig. 2.11 middle);

(ii) a buckling of fibrils parallel to the fibre axis. Since the shape of the indentation must conform with that of the diamond while the load is applied, the anisotropy observed must arise instantly upon load removal because of greater elastic recovery of the fibrillar network with strained tie molecules along the fibre axis where the stresses were largest (Fig. 2.11(*c*)).

It is useful to define the indentation anisotropy as (Rueda *et al.*, 1982)

$$\Delta H = 1 - \left(\frac{d_\parallel}{d_\perp}\right)^2 \tag{2.6}$$

where $d_\parallel$ and $d_\perp$ are the indentation diagonals parallel and perpendicular to the orientation direction respectively. Microindentation anisotropy arises for values of $\lambda \gtrsim 8$ (Fig 2.9(*a*)) and thereafter it increases linearly with draw ratio in the range investigated (Fig. 2.9(*b*)). This linear increase has been correlated to the relative increasing number of tie-taut molecules in the fibrous structure of oriented PE (Baltá Calleja, 1976).

2.7.2   Anisotropy behaviour of oriented polymers

From the foregoing it is apparent that indentation anisotropy is a consequence of high molecular orientation within highly oriented fibrils and microfibrils coupled with a preferential local elastic recovery of these rigid structures. We wish to show next that the influence of crystal thickness on ΔH is negligible. The latter quantity is independent on crystal thickness and is only correlated to the number of tie molecules and intercrystalline bridges of the oriented molecular network.

Hydrostatically annealed oriented PE

Annealing drawn PE hydrostatically at high pressure generates a wide spectrum of crystal thicknesses varying from the common oriented chain-folded structures to the chain-extended ones in which folds and ties tend to disappear (Bassett &

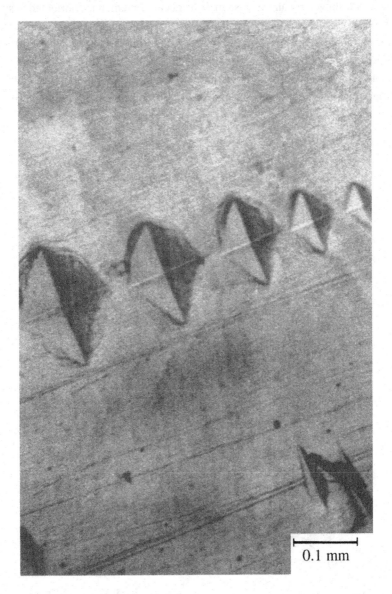

0.1 mm

Figure 2.10. Vickers indentations of oriented chain-extended OE (CEPE) along the fibre axis for various loads showing the typical anisotropic impressions. (From Baltá Calleja, 1985.)

Carder, 1973a,b). This range of crystal thicknesses, coupled with the chain axis orientation, offers a suitable model in which one can clearly distinguish between the effects of crystal extension (crystal thickness) and the influence of tie molecules on the indentation mechanism. The original material was PE ($M_w \sim 17\,000$) cold-drawn to a draw ratio $\lambda \sim 8$. Once the initial stiffening due to cold drawing is lost, increasing the chain length of the oriented lamellae varies the morphology. As expected, the increase in crystal thickness from chain-folded to chain-extended structures is followed by an increase in H as occurs in lamellar unoriented systems. Most revealing, however, is the variation of ΔH with increasing chain extension, ℓ_c of oriented lamellae (Fig. 2.12), showing first an increase up to $\ell_c \sim 500$ Å and then a decrease. Such a behaviour parallels that of the elastic modulus which, for the same chain-extension range rises to a peak and then falls off rapidly (Bassett & Carder, 1973b). Annealing at above 240 °C embrittles the polymer because

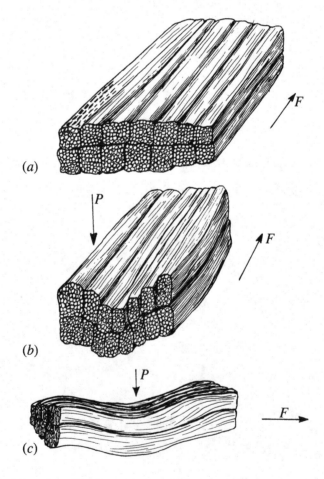

(a)

(b)

(c)

Figure 2.11. Schematics of the fibrous structure of a polymer: (a) before deformation and (b), (c) after application of a load P (see text). F: fibre axis.

of segregation of shorter molecules between chain-extended lamellae leading to fracture. The reduction in the number of tie interlamellar connections, despite the high crystal thickness achieved, contributes to the observed lowering in the overall elastic recovery of the material along the fibre axis. These experiments support the assumption that tie molecules interconnecting the oriented lamellae are not only the main contributor to the longitudinal modulus but also the rigid elements yielding to elastic recovery which causes the observed anisotropy of indentation.

Ultra-oriented solid-state extruded fibres

The influence of extrusion conditions on the microhardness of ultra-oriented PE fibres prepared by solid-state extrusion (Southern & Porter, 1970a) is a further example illustrating the value of the anisotropy measurements. This process was developed to achieve polymers of high tensile moduli (Southern & Porter, 1970b) and is based on reaching cooperative continuity for oriented high-strength bonds of a multitude of parallel polymer chains. The extended high-density PE filaments show unusually high crystal orientation (Desper et al., 1970) ($\sim$0.996), chain extension axial modulus (67 GN m^{-2}) and ultimate tensile strength (0.48 GN m^{-2}) (Capiati & Porter, 1975). The microhardening of the fibres is an increasing function of both extrusion temperature and extrusion pressure and can be explained in terms of a simultaneous improvement in the strength and lateral packing of fibrils and microfibrils (Baltá Calleja et al., 1980b). In the range of extrusion temperatures (90–137 °C) and extrusion pressures (0.24–0.5 GN m^{-2}) investigated, the micro-hardness anisotropy of the fibre surface is shown to be a unique linear function of λ (Fig. 2.13). It is independent of both the extrusion temperature and pressure. ΔH

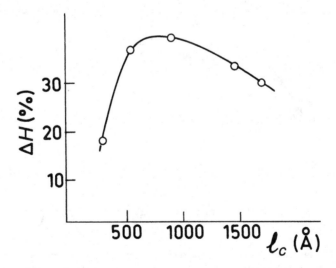

Figure 2.12. Microindentation anisotropy ΔH of oriented CEPE as a function of lamellar thickness, ℓ_c. (From Baltá Calleja, 1985.)

in the core of cleaved extruded fibres shows constant values which are independent of λ and equal the extrapolated value at the surface for the maximum attainable value of λ. These results have been explained in terms of the existing morphological differences between the outer sheath and the inner core of the fibres (Baltá Calleja *et al.*, 1980b). The inner microfibrils show the highest anisotropy because they are mainly chain-extended. At the fibre surface ΔH increases because the number of molecular connections gradually increases with λ. Microindentation anisotropy measurements can thus detect some of the fine details of the microstructure of these highly oriented fibres.

Die-drawn polymers

The linear dependence of ΔH on λ has been equally verified for ultra-oriented PE and polyoxymethylene produced by drawing the material through a heated conical die (Rueda *et al.*, 1984) (Fig. 2.13). In this case the actual draw ratio achieved depends on the drawing speed. Die-drawing has been developed from the solid-state extrusion technique, by applying a pulling force on the billet at the exit side of a converging die and by discarding the pressure applied on the billet entering the die (Coates & Ward, 1979; Gibson & Ward, 1980). It is interesting to note that the rate of ΔH increase with λ (improved recovery behaviour of the oriented molecular network) is larger for the die-drawn than for the solid-state extruded filaments. Thus for a given draw ratio and $\lambda > 12$, the former method seems to furnish products with superior elastic properties than the latter one.

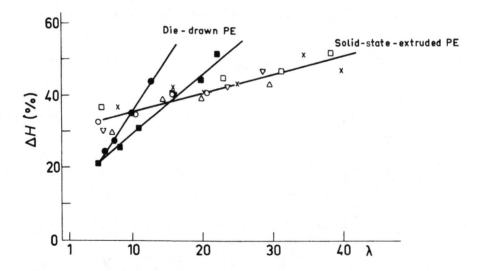

Figure 2.13. Influence of processing mode on anisotropy ΔH behaviour. Comparative correlation between ΔH and draw ratio for: solid-state ram-extruded PE at 0.24 GN m^{-2}: (O), 90 °C; ($\triangle$), 120 °C; ($\square$), 134 °C; ($\triangledown$), 137 °C; and ($\times$), 0.5 MN m^{-2}, 120 °C. Die-drawn PE with different die diameters: 4 mm ($\blacksquare$) and 15.5 mm ($\bullet$). (After Baltá Calleja, 1985.)

Injection moulded materials

Finally, some comments should be made on the anisotropy behaviour of injection moulded materials. It is known that process variables in injection moulding induce appreciable changes in the morphology of the processed material. The mechanical properties depend, in fact, upon factors such as melt temperature, ageing, mould packing and thickness (Walker & Martin, 1966). Specifically, the processing conditions of injection moulded PE markedly influence properties, such as elastic modulus, yield stress and ultimate breaking strength (Bayer, 1981). It is further known that molecular weight is a factor which strongly influences the elastic properties of the melt. The latter, in turn, control the structure of the entangled molecular network pre-existing in the molten polymer (Desper *et al.*, 1970).

The structure of the molecular network in the melt is transferred on cooling to the solid state (Picot *et al.*, 1977). On the other hand, the elasticity of the melt can be changed during injection moulding by adequately varying the molecular weight and temperature of the molten polymer. Microindentation anisotropy has been thus applied to investigate changes in microstructure and crystalline orientation of injection moulded PE using a wide range of melt temperatures and two moulding grades of different molecular weights (Rueda *et al.*, 1981). The above results favour a direct correlation between ΔH and the structure of a molecular network comprising in the solid state both crystalline and non-crystalline regions (see Chapter 7).

2.8 Microindentation *vs* nanoindentation

Nanoindentation instruments emerged as a consequence of the need to characterize mechanically the surfaces of thin films and near surfaces (within 1000 nm of the surface). The type of instrumentation and data processing needed for nanoindentation is essentially different from the microhardness tester (Pollock, 1992).

Indentation testing becomes nanoindentation testing when the size of the indent is too small to be accurately resolved by light microscopy. Nanoindentation techniques are used not only to measure microhardness and elasticity but also creep and even friction and film stress. In practice, nanoindentation testing usually involves continuous load–displacement recording as the indenter is driven into and withdrawn from the film, in contrast to single-valued measurements of contact area after load removal, which are usual with microindentation testing. Figure 2.14 illustrates the typical complete loading–unloading cycle for an amorphous PET sample. Continuous load–displacement recording, also called continuous depth recording (CDR), has proven to provide valuable additional information to conventional microindentation measurements (Polanyi *et al.*, 1988).

The advantages of CDR include a high level of precision, ease of digitization, automation and data processing. It is ideal for the measurement of creep, as well as of plastic and elastic work. However, as a sole method of measuring indent size, its disadvantage is the need for simplifying assumptions in order to: (i) separate plastic from elastic effects; (ii) determine the true zero of the depth measurements; (iii) allow for piling-up or sinking-in of material around the indent; (iv) allow for geometric imperfection of the indenter when deriving absolute microhardness values.

Nanoindentation testing by CDR does not give values of absolute microhardness directly. This is because microhardness is usually defined as load divided by indent area projected onto the plane of the surface, and this area is not explicitly measured in nanoindentation testing. However, the data can be processed on the basis of well established assumptions (Loubet *et al.*, 1984) to yield relatively direct information that is of value in quality control.

Direct information on elastic recovery, relative hardness, work of indentation, and strain rate–stress relationship (Fig. 2.15) can provide a comprehensive 'fingerprint' of a particular sample resulting, for example, from a change in either a production process or a wear test procedure. It is ideally suited to the comparison of one sample with a control or reference. The wider assumptions that are needed to derive indirect

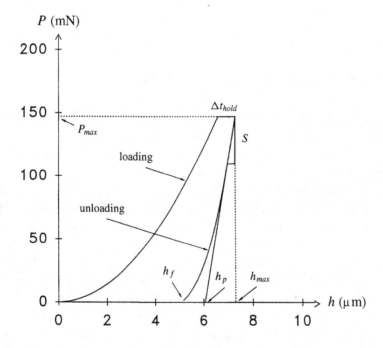

Figure 2.14. Loading–unloading cycle (force P *vs* penetration depth h) for amorphous PET (loading rate in 0.28 mN s^{-1}; holding time at peak load is 6 s) (After Flores & Baltá Calleja, 1998.)

information on the material properties that are of particular scientific value are also
included in Fig. 2.15 (Pollock, 1992).

2.8.1   Microindentation instruments

The basic principles of microindentation have been presented in Sections 2.1, 2.2
and 2.3. There is a wide variety of traditional commercial instruments which give an
accurate measurement of the microhardness of a polymer surface. In what follows
we will just summarize the main features of a typical microindentation instrument:

(a) *Loading-unit*: usually a load range between 0.1 mN and 2000 mN is available.

(b) *Indenter*: Vickers, Knoop or Berkovitch (triangle) indenters are currently used.
    The indenter is fitted on a microscope revolver and is usually designed to
    penetrate at right angles to the sample.

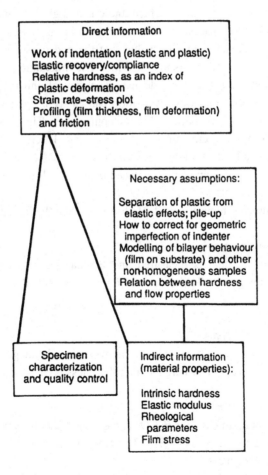

Figure 2.15. Types of information obtained from nanoindentation testing by CDR. (From
Pollock, 1992.)

(c) *X–Y stage*: this stage on which the sample is placed can be shifted in both the X and Y directions in steps of 0.01 mm over a range of 20–30 mm.

(d) *Measuring objective*: the objective of an optical microscope (usually 40×) is fitted on to the revolver and used for measurement of the indentation diagonal under a magnification of 400×. The working distance of such an objective is approximately 500 μm.

(e) *Load-time preset*: loading time can be preset or measured with a clock.

(f) *Indentation measurement*: a measuring mechanism of the ocular micrometer allows measurements in the field of the indentation image.

In modern instruments the test parameters are entered through a keyboard or through a serial interface using a compatible personal computer. At the start of the measurement, the diamond tip approaches the surface at a defined speed. When the indenter contacts the specimen surface, there is a slight load increase which is registered by the control processor. The actual load is increased according to the load gradient until it reaches the full value. At the end of the loading time, the diamond indenter is retracted to its rest position. The indentation is then measured by an integrated diagonal measuring device. The control unit reads the data and calculates the hardness. Statistical evaluation can easily be performed, since the data from a series of tests can be stored in the control unit. The sensor corresponds in shape and size to the objective of a light microscope. It screws directly into one of the objective threads on the nosepiece and can be used on upright or inverted microscopes. The diamond tip can be replaced *in situ* to allow easy interchange of Vickers and Knoop indenters.

Nowadays various systems are available for measuring the indentation diagonal:

The standard measuring eyepiece
The length of the diagonal is measured and manually entered into the control unit through the keyboard. The microhardness is then calculated and displayed automatically.

The digital measuring eyepiece
The eyepiece determines the diagonals on the microscope's image. The values are transmitted to the control unit, where the microhardness is calculated and displayed.

The video measuring system
A video channel counting detector (CCD) camera transmits the image to the monitor where the indentation is displayed. The control unit reads the diagonals and calculates the microhardness.

The integrated diagonal measuring system feeds all measured data directly to the control unit, where the results are displayed. Results can be printed on an external printer, too. The load range varies between 0.0005 and 2 N; the load gradient ranges between 0.001 and 1 N s^{-1} and the loading time spans between 0 and 10.000 s.

For microhardness investigations with very low indentation loads and extremely high local resolution (in the micrometre range) there are also instruments designed for use with commercially available scanning electron microscopes (SEM). The indentation unit is mounted on a flange of the SEM chamber like other analytical tools such as energy dispersive X-ray analysis (EDX).

2.8.2 Nanoindentation instruments

Scanning electron microscopy

Where the prime requirement is to obtain absolute values of microhardness in the sense of resistance to plastic deformation, a logical approach is to replace the optical microscope of a microindenter by an electron microscope. A nanoindentation attachment that can be used inside a SEM has been the basis of patents and is commercially available (Bangert & Wagendistrel, 1985). In principle, this approach makes it possible to establish a reliable comparison between nanoindentation microhardness values and established scales of microhardness numbers, such as those defined in national standards specifications. It is necessary to overcome the difficulties of imaging small indentations with sufficient contrast, and, at the smallest depths, to correct for the deformation of the required conductive layer of soft metal (Wagendristel *et al.*, 1987).

CDR instruments

In practice, several investigators have found the advantages of CDR to be overriding. Pollock (1992) offered a very valuable review dealing primarily with nanoindentation testing using CDR.

Vickers or Knoop indenters are sometimes used (Polanyi *et al.*, 1988; Bhushan *et al.*, 1988), but a three-sided pyramid is more common, because this shape can achieve a better approximation to a perfectly pointed apex, either by polishing or by ion erosion. The apical angles can be chosen so that the nominal relation between indent area and depth is the same as for the Vickers shape. Alternatively, a more acute angle (such as 90° between edges) can be chosen, on the grounds that thinner coatings can be tested without the data being significantly affected by the properties of the substrate. Also, sharp indenters give more consistent results when the specimen surface is rough (Newey *et al.*, 1982). Nanoindentation devices either impose a load and measure the displacement produced ('soft' testing) or measure the load needed to produce a given displacement ('hard' testing).

With the former machines (Pethica & Oliver, 1982; Wierenga & Franken, 1984), the indenter and the loading device must be mounted on a frictionless suspension. This often involves an elastic design (spring hinges and leaf springs). In one instrument (Wierenga & Franken, 1984), an air bearing is used instead, and there is no need for the weight of the indenter to be counterbalanced. To minimize kinetic and impact affects, the moment of inertia of the moving assembly should be as small as possible. The relative movement of indenter and specimen (indentation depth) is measured by means of a displacement transducer, which can be a capacitance gauge, a variable mutual inductance, or a fibre-optic device. The specifications of one commercial instrument of Micro Materials Limited (Fig. 2.16) are listed in Table 2.2.

With the 'hard' testing machines (Bhushan et al., 1988; Tusakamoto et al., 1987; Wu et al., 1988), the indentation depth is controlled, for example, by means of a piezoelectric actuator. Force transducers used in existing designs include: a load cell with a range from a few tens of micronewtons to 2 N (Wu et al., 1988; Wu 1991); a digital electrobalance with a resolution of 0.1 μN, and a maximum of 0.3 N; and a linear spring whose extension is measured by polarization interferometry.

It is possible to vary the load or, in hard machines, the displacement, either in ramp mode or with a discontinuous increment (step mode). The ramp function needs to be smooth, as well as linear, and there is evidence (Mayo & Nix, 1988) that if the ramp is digitally controlled, the data will vary for the same mean loading rate according to the size of the digitally produced load increments, unless these are very small.

Table 2.2. *Commercial nanoindentation instrument specifications. (From Pollock, 1992.)*

Ultra-microhardness measurement	
Typical depth resolution	<1 nm
X resolution/travel	0.02 μm/50 mm
Y resolution/travel	0.02 μm/50 mm
Z resolution/travel	0.02 μm/50 mm
Repositioning precision	0.1 μm
Typical force resolution	20 μN
Maximum load	0.5 N
Software	Preprogrammed control and data analysis package with menu-driven facilities for custom program development
Dimensions	500 × 400 × 300 mm
Stress measurement	
Maximum sample size	200 mm diameter
Analysis field	50 × 50 mm square
Data obtained	Film thickness, film and substrate elastic modulus, substrate shape

The basic requirements include a system for data logging and processing. Scatter of the data tends to be greater with nanoindentation than with microindentation, partly as a result of unavoidable surface roughness, but principally because the specimen volume being sampled in a single indentation is often small compared with inhomogeneities in the specimen (such as grain size or mean separation between inclusions). Thus, unless each indent is to be located at a particular site, it is usually necessary to make perhaps five, ten, or more tests, and to average the data.

As in the case of microindentation, the spacing between nanoindents must be large enough for each set of data to be unaffected by deformation resulting from

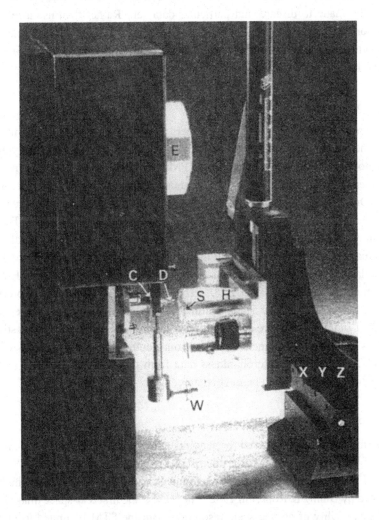

Figure 2.16. Nanoindentation instrument with CDR; XYZ, three-dimensional specimen micromanipulator; H, removable specimen holder; S, specimen; D, diamond indenter; W, balance weight for indenter assembly; E, electromagnet (load application); C, capacitor (depth transducer). (From Pollock, 1992.)

nearby indents, and the total span should be at least one or two orders of magnitude greater than the size of the specimen inhomogeneities whose effect is to be minimized by averaging. If the test results turn out to be grouped in a way that reveals differences between phases or grains, then each group should be averaged separately. In either case, the number of data points to be processed is large (Pollock, 1992).

A real-time display helps the operator to monitor the data for consistency between indents and for any systematic trend that arises, for example, from a change in the effective geometry of the indenter, if traces of material from the specimen become transferred to it. The most common reason for an inconsistent set of data is a vibration transient, the effect of which is visible at the time. A subjective decision can then be made to discard that particular data set. Rather than use a real-time display for this purpose, a more reliable approach is to use the output signal from a stylus vibration monitor (a simple modification of the detection system itself) to abort any individual test during which the vibration exceeds a certain level.

Applications to polymers

So far there have been few studies of polymers using the load–displacement method. The hardness and the elastic modulus values of polyisobutadiene rubber, poly(ether–ether–ketone), poly(methyl methacrylate) (PMMA) and Nylon-6 have been reported (Briscoe et al., 1996). In all cases, the indentation depth lay within the micrometre regime. One paper (Briscoe & Sebastian, 1996), carefully examined the influence of the indenter geometry on the microhardness and Young's modulus values of PMMA. It was shown, in addition, that the values of microhardness measured directly from the residual image correlate well with those computed using the CDR method. The plastic, elastic and flow properties of amorphous and uniaxially drawn PET have also been determined (Ion et al., 1990). The indentation depths of the impressions produced ranged between 0.5 and 5 μm. The viscoelastic–plastic properties of glassy PET in the micrometre and submicrometre range have also been investigated by means of load–displacement analysis from depth-sensing experiments (Flores & Baltá Calleja, 1998). Microhardness data from the depth-sensing and imaging methods have been shown to be in good agreement.

Surface force microscopy

Another modern instrument used for nanoindentation and for nanotribological studies of polymers is the surface force microscope (SFM) (Overney, 1995a).

The SFM is derived from the scanning tunnelling microscope (STM), which was introduced by Binnig & Rohrer in 1982. The STM is the first real-space imaging tool with the capability of atomic-scale resolution. But the STM is limited to imaging conducting surfaces. In 1986, Binnig et al., developed the SFM, which is capable of imaging both conductive and insulating surfaces. A year later, the SFM was used to measure lateral forces that occurred during sliding contact (Mate et al., 1987).

SFM is a simple but very efficient tool for studying surfaces on the submicrometre scale. The principle on which it works is very similar to profilometry, where a hard tip is scanned across the surface and its vertical movements monitored. As a result of the miniature size of the SFM tip, which is mounted at the bottom end of a cantilever-like spring, it is possible to image the corrugation of the surface potential of the sample (Overney *et al.*, 1994).

After these initial and promising lateral-force results, the friction force microscopy (FFM) was introduced. The FFM is a modified SFM with a four-quadrant photodiode, based on the laser beam deflection technique (Meyer & Amer 1988) (Fig. 2.17). The beam is emitted by a low-voltage laser diode and reflected from the rear side of the cantilever to the four-quadrant photodiode. With this detection scheme, normal and torsional forces can be measured simultaneously. The torsional forces correspond to the lateral forces measured with the instrument of Mate *et al.* (1987). In 1993, Overney introduced the threefold measurement of topography, friction and elasticity on a polymer sample using an FFM. With this latest achievement, a wide spectrum of tribological information was opened up, limited only by the lattice parameters of the sample.

The SFM has been used in studies of polymer surfaces. Theoretical calculations (Abraham & Batra, 1989; Overney, 1993) show that the SFM tip, operating in contact mode, can cause a significant distortion of the electronic and atomic structure of the measured materials. The SFM tip is therefore well suited as a micromechanical

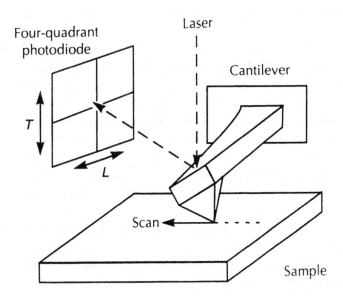

Figure 2.17. Sketch of a friction force microscope (FFM) with a beam-deflection detection scheme. Cantilever movements are monitored by a laser beam with a four-quadrant photodiode. The topography (T) is measured simultaneously with the lateral forces (L). Irreversible lateral forces are by definition frictional forces. (From Overney, 1995.)

tool. A modified SFM has been used in indentation and scratch tests on polycarbonate (PC), PMMA and epoxy (EP) surfaces (Hamada & Kaneko, 1992). With a sharp diamond tip and a penetration load of 500×10^{-9} N, indentations 50 nm wide were created, and their profiles subsequently analysed with the same instrument at a lower load of 50×10^{-9} N. Comparison with a conventional Vickers microhardness tester showed significant differences. In the Vickers microhardness test, PC and PMMA showed similar indentation marks, which were about a third larger than in EP. In the SFM test, the indentation depth of PMMA was comparable to that for EP, and half that for PC. Hamada & Kaneko (1992) concluded that the discrepancies between these two tests are due to differences in the operating regime. As compared to the SFM, the Vickers microhardness tester evidently operates with a much larger volume of material, and its applied load is ten orders of magnitude higher than that of the SFM tester. Therefore, the Vickers microhardness test is more sensitive to the bulk properties of the material, and the SFM test primarily probes the mechanical properties of the surface.

Similar results were reported by another group from scratching experiments on PC (Jung et al., 1992). With Si_3N_4 cantilevers and applied loads of 10^{-7} N, line structures were formed with an indentation depth of 10 nm and width of 70 nm. Jung et al. (1992) calculated from this experiment an average indentation pressure of the order of 10^7 N m^{-2} and compared it to the bulk compressive strength of PC, which is about 90 N m^{-2}. Since an indentation pressure of the order of 10^7 N m^{-2} is not expected on the macroscopic scale with an indentation depth of only 10 nm, it is concluded that microhardness at small loads must be much higher than measured with macroscopic testers.

With this observation the term 'surface microhardness' has been introduced for the new material property. Its technical importance is obvious for micromotors and micromachining. Future experiments and theoretical developments will show the influence of surface microhardness on macroscopic quantities. The origin of surface microhardness is still an unsolved problem. The key lies in a better understanding of the surface microhardness on the molecular scale and its relation to surface tension. It has been suggested that the difference between bulk and surface microhardness is probably due to a change of the network structure at the surface and/or the capability of the surface to reconstruct faster than the bulk to adjust to external changes (Overney, 1995a).

Let us consider one final example: the application of atomic force microscopy (AFM) relating to nanoscale scratch and indentation tests on short carbon-fibre-reinforced PEEK/polytetrafluoroethylene (PTFE) composite blends (Han et al., 1999). In the scratch test, the tip was moved across the surface at constant velocity and fixed applied force to produce grooves with nanometre scale dimensions on the PEEK matrix surfaces. The grooves consisted of a central trough with pile-ups on each side. These grooves provide information about the deformation mechanisms and scratch resistance of the individual phases. In the nanoscale, indentation and scratch of the polymeric phase involve microploughing and microcutting wear mechanisms. The harder phases, i.e. graphite and carbon fibres, are worn by microcracking events (Han *et al.*, 1998).

2.9 Outlook

In conclusion, microhardness can be seen as a bridging parameter between microstructure and macroscopic mechanical properties. Microhardness measurement offers, in addition, future possibilities for the mechanical characterization of specific parts of processed polymers on a micrometre scale (Bowman & Bevis, 1977) (prepared by extrusion, injection moulding, etc.). So far we have only considered simple polymers. Work is under way to examine more accurately the microindentation hardness of carbon-fibre-reinforced epoxy and thermoplastic composites. These composite materials represent a bridge in microhardness between pure polymers and hard metals. The localized nature of the microhardness test allows information to be obtained regarding the heterogeneity of these composites that is often not available with other analytical techniques. Relatively little work has been done, so far, on the mechanical properties of polymer surfaces at nanometre resolution (Ion *et al.*, 1990). Nanoindentation techniques (Pollock, 1992) for probing the mechanical behaviour of the top few tens of nanometres of bulk and thin polymer film specimens thus opens up future possibilities for investigating the elastic and plastic flow properties of the near-surface region of polymer materials.

Nanoindentation with CDR is being increasingly used in the characterization of submicrometre layers, surface treatments and fine particles. Modified instruments are also used to measure film stress, thickness, adhesion, scratch hardness and microfriction. Currently, the technique is best suited to providing a comprehensive quantitative fingerprint of a specific sample and to comparing it with a control or reference. This direct information includes the work of indentation, the relative hardness, the elastic compliance and the strain-rate–stress characteristics. There is no universally accepted absolute microhardness scale that applies to nanoindentation. With the help of a number of assumptions, it is possible to derive values of intrinsic material properties, such as microhardness or modulus, although it is not yet clear to what extent these values depend on test variables, such as the geometry of the indenter used. Furthermore, the effect of variations in loading rate, for example, tends to be especially noticeable at the submicrometre scale.

A survey of the nanoindentation literature suggests that technical advances are likely to emphasize the following points as formulated by Pollock (1992):

■ The analysis of indentation creep data as an important aspect of material characterization.

- The introduction of a new hardness scale (Field, 1988) based on the CDR method.

- The development of multipurpose nanoindentation instruments that also perform scratch testing, profiling, and measurements of scratch hardness, film stress, friction, and other surface-mechanical properties.

- Use in routine industrial testing, which requires improved automation so that the monitoring of changes in indenter shape, for example, is more reliably performed.

- The introduction of specimen heating stages, together with a satisfactory method of compensating for thermal drift.

In the longer term, picoindentation instruments are likely to be widely used to extend the technique to a still smaller scale, with the help of techniques developed for atomic force microscopy. Already, plastic deformation at depths of a few atomic layers, as well as the effect of surface forces, have been quantified by means of depth–load measurements, using a point force microscope, i.e. an AFM operated in static (non-scanning) mode (Burnham & Colton, 1989).

Our understanding of polymer surfaces will be much improved by performing experiments on the nanometre scale. Computer simulations and models can easily be confirmed by experiments on such small scales (Overney et al., 1993, 1994; Noether & Whitney, 1973). In the future, new concepts, which find analogues in statistical mechanics and thermodynamics, may be developed because of the small contact area. One of the important future goals of the SFM and FFM techniques is to connect phenomenological aspects in tribology with a molecular understanding. To achieve such a goal scientists in nanotribology will have to follow atomistic and statistical principles, avoid empirical formulae and search for conformity between experimental data and atomistic theoretical models and computer simulations.

2.10   References

Abraham, F.F. & Batra, I.P. (1989) *Surf. Sci.*, **209**, L125.

Baer, E., Maier, R.E. & Peterson, R.N. (1961) *SPE Journal* November, 1203.

Baltá Calleja, F.J. (1976) *Colloid & Polymer Sci.* **254**, 258.

Baltá Calleja, F.J. (1985) *Adv. Polym. Sci.* **66**, 117.

Baltá Calleja, F.J. & Bassett, D.C. (1977) *J. Polym. Sci.* **58C**, 157.

Baltá Calleja, F.J. & Peterlin, A. (1970) *J. Macromol. Sci. Phys.* **B4**, 519.

Baltá Calleja, F.J., Flores, A., Ania, F. & Bassett, D.C. (2000) *J. Mater. Sci.* **35**, 000.

Baltá Calleja, F.J., Martínez-Salazar, J., Cackovic, H. & Loboda-Cackovic, J. (1981) *J. Mater. Sci.* **16**, 739.

Baltá Calleja, Mead, W.T. & D.R., Porter, R.S. (1980a) *Polym. Eng. Sci.* **20**, 39.

Baltá Calleja, F.J., Peterlin, A. & Crist, B. (1972) *J. Polym. Sci.* **A210**, 1749.

Baltá Calleja, F.J., Rueda, D.R., Porter, R.S. & Mead, W.T. (1980b) *J. Mater. Sci.* **15**, 765.

Baltá Calleja, F.J., Santa Cruz, C., Asano, T. (1992) *J. Polym. Sci. Polym. Phys. Ed.* **31**, 557.

Baltá Calleja, F.J., Santa Cruz, C., González Arche, A. & López Cabarcos, E. (1992) *J. Mater. Sci.* **27**, 2124.

Bangert, H. & Wagendristel, A. (1985) *Rev. Sci. Instrum.* **56**, 1568

Bangert, H., Wagendrizted, A. & Aschinger, H. (1983) *Philips Electron Optics Bull.* **119**, 17.

Bassett, D.C. & Carder, D.R. (1973a) *Phil. Mag.* **28**, 513.

Bassett, D.C. & Carder, D.R. (1973b) *Phil. Mag.* **28**. 535.

Bayer, R.K. (1981) *Colloid & Polym. Sci.* **259**, 303.

Bhushan, B., Williams, V.S. & Shack, R.V. (1988) *Trans. ASME J. Tribol.*, **110**, 563.

Binnig, G. & Rohrer, H. (1982) *Helv. Phys. Acta* **55**, 726.

Binnig, G., Quate, C.F. & Gerber, C. (1986) *Phys. Rev. Lett.* **56**, 930.

Bowmann, J. & Bevis, M. (1977) *Colloid & Polymer Sci.* **255**, 954.

Briscoe, B.J. & Sebastian, K.S. (1996) *Proc. R. Soc. A* **452**, 439.

Briscoe, B.J., Sebastian, K.S. & Sinha, S.K. (1996) *Phil. Mag.* **74**, 1159.

Burnham, N.A. & Colton, R.J. (1989) *J. Vac. Sci. Technol.* **A7**, 2906.

Capiati, N.J. & Porter, R.S. (1975) *J. Polym. Sci. Polym. Phys. Ed.* **13**, 1177.

Coates, P.D. & Ward, I.M. (1979) *Polymer* **20**, 1553.

Desper, C.R., Southern, J.H., Ulrich, R.D. & Porter, R.S. (1970) *J. Appl. Phys.* **41**, 4284.

Eyerer, P. & Lang, G. (1972) *Kunststoffe* **62**, 222.

Field, J.S. (1988) *Surf. Coat. Technol.* **36**, 817.

Flores, A. & Baltá Calleja, F.J. (1998) *Phil. Mag.* **78**, 1283.

Gibson, A.G. & Ward, I.M. (1980) *J. Mater. Sci.* **15**, 979.

Grubb, D.T. & Keller, A. (1980) *J. Polym. Sci. Polym. Phys. Ed.* **18**, 207.

Hamada, R. & Kaneko, R. (1992) *Ultramicroscopy* **42–44**, 184.

Han, Y., Schmitt, S. & Friedrich, K. (1998) *Tribology Intern.* **31**, 715.

Han, Y., Schmitt, S. & Friedrich, K. (1999) *Appl. Composite Mater.* **6**, 1.

Hoffman, J.D., Davis, G.T. & Lauritzen Jr J.J. (1976) *Treatise on Solid State Chemistry*, Vol. 3 (Hannay, N.B. ed.) Plenum Press, New York, Chapter 7.

Ion, R.H., Pollock, H.M. & Roques-Carmes, C. (1990) *J. Mater. Sci.* **25**, 1444.

Jung, T.A. *et al.* (1992) *Ultramicroscopy* **42–44**, 1446.

Kiho, H., Peterlin, A. & Geil, P.H. (1964) *J. Appl. Phys.* **35**, 1599.

Kleinhärteprüfer Durimet, Ernst Leitz GmbH, Wetzlar, Liste 72–5a, 1964.

Langford, G. (1984) *Encyclopedia of Chemical Technology*, third edition, (Standen, A. & Kik-Othmer eds.) Wiley-Interscience Publ., New York, p. 118.

Lawn, B.R. & Howes, V.R. (1981) *J. Mater. Sci.* **16**, 2745.

Loubet, J.L., Georges, J.-M., Marchesini, O. & Meille, G. (1984) *Trans. ASME J. Tribol.* **106**, 43.

Love, A.E.M. (1927) *The Mathematical Theory of Elasticity*, fourth edition, Dover Publications, London, p. 183.

Marsh, D.M. (1964) *Proc. Roy. Soc. London* **A279**, 420.

Martin, B., Pereña, J.M., Pastor, J.M. & De Saja, J.A. (1986) *J. Mater. Sci. Lett.* **5** 1027.

Mate, C.M., McClelland, G.M., Erlandsson, R. & Chiang, S. (1987) *Phys. Rev. Lett.* **59**, 1942.

Maxwell, B. (1955) *Modern Plastics* **32**, 125.

Mayo, M.J. & Nix, W.D. (1988) *Measuring and Understanding Strain Rate-Sensitive Deformation with the Nanoindenter, Strength of Metals and Alloys*, ICSMA 8 (Kettunen, P.O., Lepisto, T.K. & Lehtonen, M.E., eds.) Pergamon, London, p. 1415.

Meinel, G., Morossoff, N. & Peterlin, A. (1970) *J. Polym. Sci.* **A2**, 1723.

Meyer, G. & Amer, N.M. (1988) *Bull. Am. Phys. Soc.* **33**, 319.

Morossoff, N. & Peterlin, A. (1970) *J. Polym. Sci.* **A2**, 1237.

Müller, K. (1970) *Kunststoffe* **60**, 265.

Newey, D., Pollock, H.M. & Wilkins, M.A. (1982) *The Ultra-Microhardness of Ion-Implanted Iron and Steel at Sub-Micron Depths and its Correlation with Wear-Resistance, Ion Implantation into Metals* (Ashworth, V. *et al.*, eds.) Pergamon, London, p. 157.

Noether, H.D. & Whitney, W. (1973) *Kolloid Z.Z. Polym.* **251**, 991.

Overney, G. (1993) *Scanning Tunneling Microscopy III* (Wiesendanger, R. & Güntherodt, H-J., eds.) Springer-Verlag, Berlin, p. 251.

Overney, R.M. (1995a) *Trends Polym. Sci.* **3**, 359.

Overney, R.M. (1995b) *Langmuir* **10**, 1281.

Overney, R.M., Takano, H., Fujihira, M., Paulus, W. & Ringsdorf, H (1994) *Phys. Rev. Lett.* **72**, 3546.

Paar, A., A.G. (1996) Ultra-microhardness tester MHT-Y, US-Patent No. 4 611 487.

Pereña, J.M., Martin, B. & Pastor, J.M. (1989) *J. Mater. Sci. Lett.* **8** 349.

Peterlin, A. (1987) *Colloid & Polym. Sci.* **265**, 357.

Peterlin, A. & Baltá Calleja, F.J. (1970) *Kolloid Z.Z. Polym.* **242**, 1093.

Pethica, J.B. & Oliver, W.C. (1982) *Ultra-Microhardness Tests on Ion-Implanted Metal Surfaces, Ion Implantation into Metals* (Ashworth, V. *et al.*, eds.) Pergamon, London, p. 373.

Pethrick, R.A. (1993) *Trends Polym. Sci.* **1**, 226.

Picot, C., Duplessix, R., Decker, D., Benoit, H., Boue, F., Cotton, J.P., Daoud, M., Farmoux, B., Jannink, G., Nierlich, M., De Vries, A.J. & Pincus, P. (1977) *Macromolecules* **10**, 436.

Polanyi, R.S., Ruff Jr, A.W. & Whitenton, E.P. (1988) *J. Test. Eval.* **16**, 12.

Pollock, H.M. (1992) *ASM Handbook*, Vol. 18: *Friction Lubrication, and Wear Technology*, ASM International Materials Park, Ohio, p. 419.

Pollock, H.M., Maugis, D. & Barquins, M. (1986) *Microindentation Techniques in Materials Science and Engineering* (Balu, P.J. & Lawn, B.R., eds.) ASTM, Philadelphia, p. 47.

Racké, H.H. & Fett, T. (1971) *Materialprüfung* **13**, 37.

Rikards, R., Flores, A., Kushnevski, V. & Baltá Calleja, F.J. (1998) *J. Computational Mater. Sci.* **II**, 233.

Rueda, D.R., Ania, F. & Baltá Calleja, F.J. (1982) *J. Mater. Sci.* **17**, 3427.

Rueda, D.R., Baltá Calleja, F.J. & Bayer, R.K. (1981) *J. Mater. Sci.* **16**, 3371.

Rueda, D.R., García, J., Baltá Calleja, F.J., Ward, I.M. & Richardson, A. (1984) *J. Mater. Sci.* **19**, 2615.

Sakaoku, K. & Peterlin, A. (1971) *J. Polym. Sci.* **A2**, 895.

Santa Cruz, C., Baltá Calleja, F.J., Asano, T. & Ward, I.M. (1993) *Phil. Mag.* **68** 209.

Southern, J.H. & Porter, R.S. (1970a) *J. Macromol. Sci. Phys.* **4**, 541.

Southern, J.H. & Porter, R.S. (1970b) *J. Appl. Polym. Sci.* **14**, 2305.

Stilwell, N.A. & Tabor, D. (1961) *Proc. Phys. Soc. London* **78**, 169.

Tabor, D. (1951) *The Hardness of Metals*, Oxford C. Press, New York.

Thomas, D.G. & Stavely, L.A.K. (1952) *J. Chem. Soc.* **1952**, 4569.

Tusakamoto, Y., Yamaguchi, H. & Yanagisawa, M. (1987) *Thin Solid Films* **154**, 171.

Wagendristel, A., Bangert, H., Cai, X. & Kaminitschek, A. (1987) *Thin Solid Films*, **154**, 199.

Walker, J.L. & Martin, E.R. (1966) *Injection Moulding of Plastics*, Plastics Institute, Iliffe Books, London.

Wierenga, P.E. & Franken, A.J.J. (1984) *J. Appl. Phys.* **55**, 4244.

Wilding, M.A. & Ward, I.M. (1978) *Polymer* **19**, 969.

Wilding, M.A. & Ward, I.M. (1981) *Polymer* **22**, 870.

Wu, T.W. (1991) *J. Mater. Res.* **6**, 407.

Wu, T.W., Hwang, C., Lo, J. & Alexopoulos, P. (1988) *Thin Solid Films* **166**, 299.

Zalwert, S. (1970) *Makromol. Chem.* **131**, 205.

Chapter 3

Microhardness of glassy polymers

3.1 Introduction to glassy polymers

3.1.1 The glass transition

When a polymer is cooled down from the liquid or rubbery state, it becomes much stiffer as it goes through a certain temperature range. This stiffening is the result of one of two possible events: crystallization or glass transition. For crystallization to occur, the polymer molecules must be sufficiently regular along their length to allow the formation of a crystalline lattice and the cooling rate must be slow enough for the crystallization process to take place before the molecular motions become too sluggish. When the polymer fails to crystallize for either reason, the amorphous, liquid-like structure of the polymer is retained, but the molecular motion becomes frozen-in and the material turns into a glass. Such a glass transition occurs over a finite temperature interval, but is still realized abruptly enough to merit the term 'transition'. The glass transition can be recognized by the change in many properties of the material, the most important one, from a practical point of view, being the increase in the modulus of the material by several orders of magnitude.

Glass formation can be achieved with many low-molecular-weight materials and with certain metallic alloys by special preparation techniques, such as rapid quenching. With polymers, the opportunities for irregularity along the chain are numerous and the crystallization rate is inherently slow. As a result, the formation of the glassy state is a more common occurrence. Thus the glassy state of polymeric materials is a state equal in importance to the semicrystalline state and the rubbery state. Also, the melting temperature T_m and the glass-transition temperature T_g are the two most important parameters of a given polymer that characterize its properties over a wide temperature range.

The volume–temperature relationship of a typical polymer is depicted in Fig. 3.1. Upon crystallization at temperatures moderately below T_m, the specific volume of the polymer decreases significantly in comparison with that of the amorphous polymer. If crystallization is prevented, either because of irregularities in the polymer chain or because of a rapid cooling rate, the material undergoes a glass transition manifested not by a change in specific volume but by a change in the thermal expansion coefficient, α. If the enthalpy of the polymer were measured by calorimetry and plotted against temperature, the resulting enthalpy–temperature relationship would be very similar to the plot in Fig. 3.1. In any first-order transition, such as melting or boiling, the volume and entropy, which are both first derivatives of the free energy, undergo a discontinuous change. In the second-order transition,

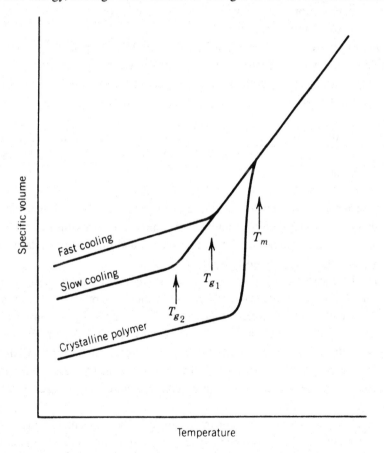

Figure 3.1. Schematic illustration of the volume–temperature relationship of a typical polymer. When the polymer is prevented from crystallization, it is brought to temperatures below T_m in an amorphous state and is then turned into a glass at the glass-transition temperature T_{g1} or T_{g2}, which depends on the cooling rate. (From Roe, 1990.)

the second derivatives of the free energy, such as α and the heat capacity, undergo a discontinuous change. A glass transition has the appearance of the second-order transition in this sense. The glass-transition phenomenon as observed ordinarily is, however, a non-equilibrium phenomenon.

The values of T_g reported for the same polymer often differ greatly, sometimes as much as 10–20 °C, because the glass transition occurs over a temperature range rather than at a single, sharply defined temperature, and because the observed T_g varies somewhat, depending on the method of measurement used and on the thermal history of the sample.

It can be stated that polymers with flexible backbone chains have low T_g values and those with stiff chains have high T_g values.

There exists a fairly good correlation between the T_m and T_g for a large number of polymers. A useful rule of thumb is that the ratio T_g/T_m is 1/2 for symmetrical polymers, i.e. those containing a main-chain atom having two identical substituents, and 2/3 for unsymmetrical polymers.

For high-molecular-weight polymers T_g is essentially independent of molecular weight, but as the polymer chain length becomes shorter the T_g is likely to decrease appreciably. This lowering of T_g is essentially an end effect: the two ends of the polymer chain are able to move about more freely than a segment in the chain interior.

The application of pressure compresses the volume of a liquid and consequently its free volume is reduced. This retards molecular mobility, and the glass transition occurs at a higher temperature.

For random copolymers and miscible polymer blends, only a single T_g, which is usually intermediate between the T_g of the corresponding neat homopolymers, is observed. For block copolymers with mutually incompatible blocks, the microdomains formed by the different blocks exhibit different T_g, and for incompatible polymer blends separate T_g values are also observed.

T_g can be measured by dilatometry, refractive index, differential scanning calorimetry, dynamic mechanical methods and by dielectric relaxation techniques.

The structure of an amorphous material is characterized by the absence of a regular three-dimensional arrangement of molecules or subunits of molecules extending over distances that are large compared to atomic dimensions, i.e. there is no long-range order. However, due to the close packing of the particles in the condensed state, there is a certain regularity of the structure on a local scale, denoted short-range order.

The structure, however, is not static but is subject to thermally driven fluctuations. The local structure changes continuously as a function of time due to orientational and translational molecular motions. The time scale of these motions may range from nanoseconds up to several hundred years. The structure of the amorphous state as well as its time-dependent fluctuations can be analysed by various scattering techniques, such as X-ray, neutron, electron and light scattering.

Although macroscopic data do not provide direct information on the structure of the amorphous state, some tentative conclusions can be drawn. The dense packing of molecules in the fluid or glassy amorphous state, as opposed to the dilute packing in the gaseous state, makes short-range positional order mandatory. This is illustrated schematically in Fig. 3.2 for the particular case of subunits of chain molecules.

3.1.2   Physical ageing

A very general and important characteristic feature of glasses is their ability to undergo ageing. Physical ageing occurs because of the inherent instability of the amorphous glassy state. The process is also known as volume recovery, structural relaxation, stabilization, annealing, etc. The process can be characterized as follows: products of glassy polymers are usually made by shaping them in the molten (rubbery) state, followed by fixation of the shape by rapid cooling to below the glass-transition temperature. During cooling, the material solidifies and stiffens, but the period is too short for this process to be completed. Solidification and stiffening continue during the service life of the product. Thus, many material properties change with time. For example, the elastic modulus and the yield stress increase, the ductility may decrease, the creep and stress relaxation rates decrease, the dielectric relaxation is delayed, and the propensity for crazing and shear banding is enhanced.

As mentioned above, physical ageing is a consequence of the instability of the glassy state and is therefore found in all glassy materials: amorphous glassy polymers as well as low-molecular-weight organic glasses, substances such as bitumen,

Figure 3.2. Short-range order in chain molecules in the condensed state. (From Voigt-Martin & Wendorff, 1990.)

cured epoxies, inorganic glasses and glassy metals. Typical aspects of ageing include its reversibility and deageing ability.

In contrast to chemical ageing or degradation, physical ageing is a reversible process, i.e. by reheating the aged material to above the glass-transition temperature T_g, the original state of thermodynamic equilibrium is recovered. A renewed cooling will induce the same ageing effects as before (Struik, 1990).

So far, the phenomena discussed pertain to isothermal ageing after a quench (rapid cooling) from above to below T_g. Under such circumstances, ageing always runs in the same direction (increasing retardation times). Peculiar phenomena occur, however, when after a period of ageing at temperature T_1, the material is heated to a final temperature T_∞ between T_1 and T_g ($T_1 \leq T_\infty \leq T_g$). A good example is poly(vinyl chloride) (PVC) with $T_g \simeq 80\,^\circ\text{C}$. Struik (1978) compared two samples for which the ageing times at $T_1 = 20\,^\circ\text{C}$ differ by a factor of 18. When tested at $20\,^\circ\text{C}$, their creep properties differed considerably, but all differences disappeared when the tests were done at $50\,^\circ\text{C}$.

The facts concerning ageing discussed so far have a number of practical consequences. Firstly, data show that ageing time t_e should be included in the specification of material properties, particularly for low-frequency properties such as creep and stress or dielectric relaxation. Secondly, deageing phenomena should be taken into account in experimental studies, i.e. annealing studies. The third consequence of ageing concerns the prediction of long-term behaviour from short-time tests. In a long-term mechanical deformation process such as creep, two processes occur simultaneously: firstly the creep process itself and secondly, the ageing process that continuously stiffens the material and delays the creep. Methods for predicting long-term behaviour that take these simultaneous effects into consideration have been worked out (Struik, 1978).

Deageing effects can also be brought about by high mechanical stresses. In the extreme cases of yielding and necking, the deageing is complete and the material behaves as if it were actually heated to above T_g.

Ageing plays a key role in the non-linear viscoelastic behaviour of polymers. When left at rest and at constant temperature, there is continuous stiffening. However, when the aged material is slightly heated or mechanically deformed, it is deaged and softened (Struik, 1978, 1983).

3.2   Temperature dependence of microhardness in polymer glasses: determination of T_g

In the present section it will be shown that microhardness can conveniently detect the glass transition temperature T_g by following H as a function of temperature. We will illustrate the temperature dependence of hardness in case of two amorphous polymers – PMMA and poly(vinyl acetate) (PVAc) – and two semicrystalline

polymers – PET and poly(aryl-ether-ether-ketone) (PEEK) quenched into the fully
amorphous state (Ania *et al.*, 1989).

It should be noted that the first pair of polymers (PMMA and PVAc) are not
capable of crystallizing and, therefore, are always present in the amorphous state,
which is not the case with the second pair. PET and PEEK are high-melting
semicrystalline polymers distinguished by relatively low crystallization rates and

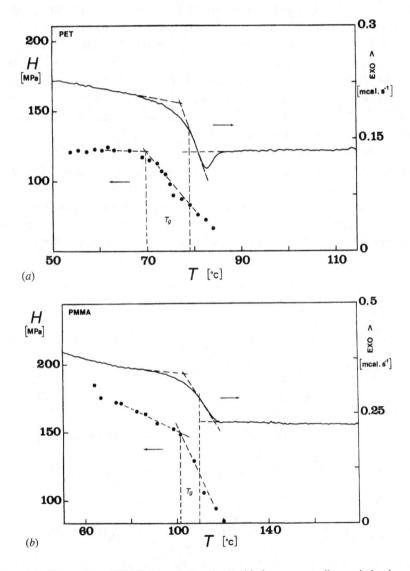

Figure 3.3. Comparison of DSC thermograms (top) with the corresponding variation in
microhardness H (solid symbols) as a function of temperature for fully amorphous (*a*) PET
and (*b*) PMMA. The vertical dashed lines denote T_g values derived from both techniques.
(From Ania *et al.*, 1989.)

this means that a fully amorphous glassy state can be obtained by rapid cooling (quenching) from the molten state. This approach has been used to prepare glasses from PET and PEEK. We will consider thin (0.5 mm) sheets of the above four materials which have been compression moulded at the following temperatures: 280 °C (PET); 400 °C (PEEK); 170 °C (PMMA); 120 °C (PVAc), and quenched well below their respective glass transition temperatures. The microhardness indentation was measured in the temperature range between room temperature and 150 °C. The glass transition temperatures for PMMA and for the amorphous and semicrystalline PET samples were also determined using a differential scanning calorimeter (DSC).

It is interesting to compare the softening of the material near T_g measured by H, as a function of temperature, with the corresponding DSC traces for amorphous PET and PMMA (Fig. 3.3). These experiments demonstrate that both methods, H and DSC, yield a similar measure of T_g. The apparent difference in the T_g values obtained ($\sim$10 °C) is a result of the fast heating rate used in the DSC determination in contrast to the quasi-static measurement in the case of H.

In order to examine the possible influence of the crystalline phase on the T_g value, the H of a PET sample with a degree of crystallinity of 0.38 was studied. Figure 3.4 compares the dependence of microhardness upon temperature for amorphous and semicrystalline PET. In order to separate the ageing contribution from the pure temperature dependence of the amorphous material the microhardness data for the former were taken after long annealing times ($t = 100$ h). The T_g value

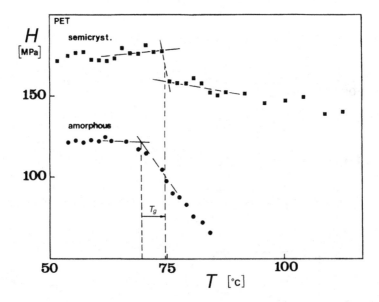

Figure 3.4. Microhardness H as a function of temperature for two PET samples: (top) crystallized and (bottom) quenched into the fully amorphous state. A shift of T_g towards higher temperatures for the crystalline polymer is observed. (From Ania *et al.*, 1989.)

for crystalline PET is clearly evidenced by a sudden H decrease, followed by a substantially slower temperature rate decrease than that obtained for the amorphous polymer. Most interesting is the fact that the T_g for the semicrystalline polymer is ~5 °C higher than for the amorphous material. This result is in accordance with the concept that crystals disturb the amorphous phase and reduce segmental mobility (Struik, 1987). As a consequence of this molecular immobilization the T_g range in the semicrystalline polymer is extended and shifted towards higher temperatures.

With respect to the influence of crystallinity (w_c) on the T_g value it is noteworthy that in earlier dilatometric measurements on the same polymer but in a wider range of w_c values (0–50%) complete dependence was observed (Fakirov, 1978; Simov et al., 1970). It was found that with the rise of w_c, T_g first increases up to 25–30%, and thereafter decreases when a crystallinity of around 50% is reached. The initial increase in T_g is actually related to the mobility-restriction effect of the numerous small crystallites (provided the low-crystallinity samples are obtained via high

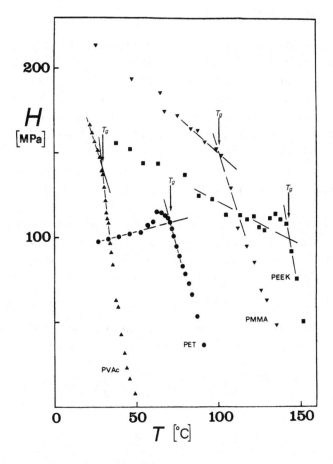

Figure 3.5. Microhardness H as a function of temperature for four amorphous polymers. Glass transition values are denoted by arrows. (From Ania et al., 1989.)

undercooling). The decrease of T_g at the highest w_c values is due to the larger mobility in the amorphous regions. This is due to the fact that the much larger crystallites are not so numerous as in the previous case (provided the high-crystallinity samples are prepared at significantly smaller undercooling) (Simov *et al.*, 1970).

Figure 3.5 shows the temperature variation of H for the four above mentioned polymers, namely PMMA, PVAc, PET and PEEK. In the case of the two amorphous polymers (PMMA and PVAc) H decreases with T and the T_g value can be clearly identified with a bend in the H *vs* T plot. However, the two semicrystalline materials quenched into the amorphous state (PET and PEEK) show an apparent maximum just before the glass transition takes place. Also, in the case of PET tested immediately after quenching it is observed that H increases with T above room temperature. These phenomena will be discussed in Section 3.3 in the light of the physical ageing undergone by the above polymers. It is seen that H follows an exponential decrease as a function of T given by (see eq. (2.5))

$$H = H_0 \exp[-\beta(T - T_0)] \tag{3.1}$$

where H_0 is the hardness measured at a given reference temperature T_0, and β is the coefficient of thermal softening. Table 3.1 lists the β coefficients for the investigated polymers, below and above T_g (once the effect of physical ageing has been minimized). It is found that for $T < T_g$, $\beta \sim (1.2-20) \times 10^{-3}$ K^{-1}, i.e. the β values are of the same order of magnitude as the data obtained for the crystalline phase in other polymers as found by Baltá Calleja (1985), Martínez-Salazar & Peña (1985) and Ania & Kilian (1986). in contrast, for $T > T_g$, H decreases at a much higher rate ($\beta = (33-144) \times 10^{-3}$ K^{-1}). As a consequence the microhardness of the amorphous phase becomes negligible within a temperature range $\Delta T \sim 50\,°C$ above T_g.

In the glassy state (below T_g) the critical stress required to plastically deform the amorphous molecular network (H) involves displacement of bundles of chain segments against the local restraints of secondary bond forces and internal rotations. The intrinsic stiffness of these polymers below T_g leads to H values which are 3–4

Table 3.1. *Glass transition temperature T_g as measured from the microhardness and coefficient of thermal softening β for the polymers investigated.*

Polymer	T_g (K)	$\beta \times 10^3$ (K^{-1})	
		$T < T_g$	$T > T_g$
PVAc	302	20.0	144
PMMA	373	4.6	33
PET	343	4.0	36
PEEK	416	4.0	56

times larger than those obtained for typical flexible polymers. At the glass transition there is an onset of liquid-like motions involving longer chain segments. These motions of larger bundles above T_g require more free volume, and lead to the faster H decrease observed (see Fig. 3.5).

A model to account for the yield behaviour of amorphous glassy polymers, making use of a dislocation analogue, has been proposed by Bowden & Raha (1973).

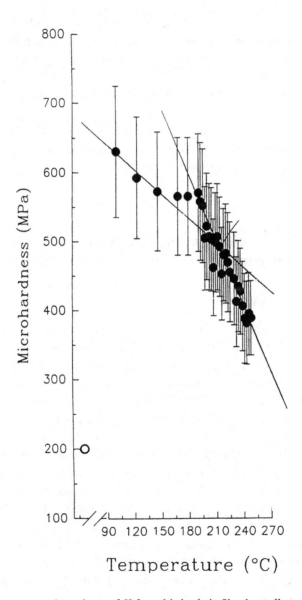

Figure 3.6. Temperature dependence of H for a dried gelatin film thermally treated at high temperature for several hours. For comparison the H value measured at room temperature is also given (open circle). (From Vassileva *et al.*, 1998.)

These authors envisaged the critical step in the yield process as being the nucleation under stress of small disc-sheared regions (analogous to dislocation loops) that form with the aid of thermal fluctuations. The model explains quantitatively the variation of the yield stress with temperature, strain rate and hydrostatic pressure, using only two parameters, the shear modulus of the material and the 'Burgers vector' of the shared region which is a constant related to the molecular dimensions of the polymer.

The temperature dependence of H for a dry gelatin film has been studied (Vassileva *et al.*, 1998) in the temperature interval 90–250 °C. The results are plotted in Fig. 3.6. A well expressed decrease in H with increasing temperature can be seen. The intercept of the two straight lines yields $T_g = 215$ °C which is in a very good agreement with the reported value of 217 °C for amorphous dry gelatin (Rose, 1987).

It can be concluded that microhardness has been shown to be a promising technique for detecting accurately the glass transition temperature of amorphous and semicrystalline polymers in addition to the techniques commonly used for this purpose (see Section 3.1).

3.3   Physical ageing of polymer glasses as revealed by microhardness

3.3.1   Homopolymers

To examine the origin of the anomalous behaviour of PET with respect to the other investigated polymers (a positive slope of microhardness H as a function of temperature below T_g, see Fig. 3.5), the microhardness for a sample quenched to a constant temperature below T_g was measured as a function of storage time (to be called the annealing time, t_a). Figure 3.7 shows the results obtained, indicating that the slope of the H vs T plot below T_g conspicuously depends on the thermal history of the samples. Beyond T_g all curves tend to similar H values with a loss of memory of the previous thermal history. This reversible phenomenon, commonly known as physical ageing, reveals the influence of the molecular relaxation in the glassy state, showing a tendency towards a more compact molecular arrangement, reducing segmental mobility and leading, as a result, to an increase of microhardness. For longer annealing times, the H vs T plot resembles that exhibited by the rest of the polymers from Fig. 3.5. The physical ageing for PET as a function of long annealing times, t_a, and thereafter annealed for a relatively short time at different temperatures T_a below T_g ($\Delta T = T_g - T_a$) has been investigated in detail. The data of Fig. 3.8 follow a time increase of H of the type

$$H = A \log t_a + K \qquad (3.2)$$

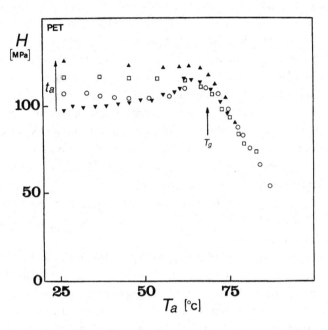

Figure 3.7. Microhardness H as a function of temperature for various PET samples for different annealing times ranging from $t_a = 0$ h (lowest curve) up to $t_a = 100$ h (highest curve). (From Ania *et al.*, 1989.)

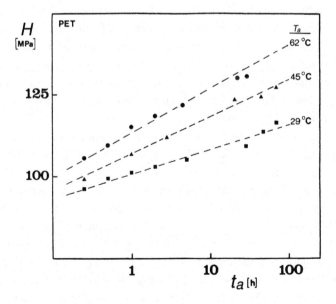

Figure 3.8. Time increase of microhardness H for PET at different annealing temperatures T_a below T_g. T_a(°C): (●) 62; (▲) 45; (■) 29. (From Ania *et al.*, 1989.)

in the range of 0–100 h, where the A and K parameters are increasing functions of T_a. The rate of microhardness increase due to physical ageing with temperature explains the apparent maximum near T_g observed in some of the investigated polymers (Fig. 3.5). The H increase with time is correlated with the well known free-volume decrease with time (Struik, 1978). This behaviour is characteristic for all types of glasses, including other non-polymeric materials. Attempts to correlate the time increase of microhardness with a possible variation in the average intermolecular distance and 'cluster size' as derived from the integral breadth of the X-ray diffraction halo do not offer a definite answer. Apparently X-ray diffraction results are not sufficiently sensitive to reveal the molecular relaxation motions which are detected by microhardness. A similar logarithmic H increase with t_a has been shown to describe the annealing behaviour of semicrystalline polymers (near T_m, the melting temperature) (Baltá Calleja, 1985). However, the molecular mechanism involved in this case is evidently quite different, mainly entailing a thickening of the crystallites (Rueda et al., 1985).

Rueda et al. (1995), have followed the physical ageing of another glassy polymer. They investigated the variation in microhardness of poly(ethylene naphthalene-2,6-dicarboxylate) (PEN) stored in an ambient atmosphere for different times and after annealing at different temperatures below T_g for different periods of time. This material has stiffer molecules than PET due to the presence of a naphthalene ring instead of the benzene ring in the backbone chain.

PEN is becoming of increasing interest as a replacement of PET because of its higher glass transition temperature of 123 °C.

Amorphous films of PEN, obtained by melt pressing at about 290 °C and quenching in ice water, were stored for different periods of time at room temperature. The physical ageing of these PEN samples was enhanced by means of annealing at different temperatures T_a below $T_g = 123$ °C, for selected annealing times t_a. The samples were annealed in an ambient atmosphere under N_2 flow using a hot-stage device at $T_a = 80, 90, 105$ and 115 °C. In this case one has to assume that water vapour is not completely removed.

In all cases H first increased as a function of annealing time t_a, approximately linearly with t_a, then passed through a maximum for a given t_a value which depends on T_a, and finally decreased towards steady values (~190 MPa) which were larger than the initial ones, as shown in Fig. 3.9. It is noteworthy that the maximum microhardness is similar for the different temperatures used (80, 90, 105 and 115 °C). However, the maximum level of microhardening observed for the heat-treated samples is very dependent on the ageing (storage time) of the films investigated. In addition, the rate of microhardening for a given temperature is larger the longer is the storage time. For all samples the limiting H values, after say 60 min of annealing, are larger (5–6 MPa) than the corresponding initial values. Finally, it was observed that a further annealing treatment of the 'dried' samples yielded no appreciable change in the hardness of the samples.

Physical ageing, which represents a change in the mechanical properties of the material, can be discussed in the light of the analysis of results deduced from the dependence of H for the initial aged PEN samples, annealed at 105 °C on the annealing time (Rueda *et al.*, 1995). The influence of the storage (ageing) on the level of microhardening at room temperature follows a logarithmic increase (long-time hardening) which is characterized by a straight line (see Fig. 3.10). Annealing below T_g (T_a = 80–115 °C) not only increases the hardening rate but also yields the maximum hardness values, H_{max}, observed for T_a = 105 °C. The values of

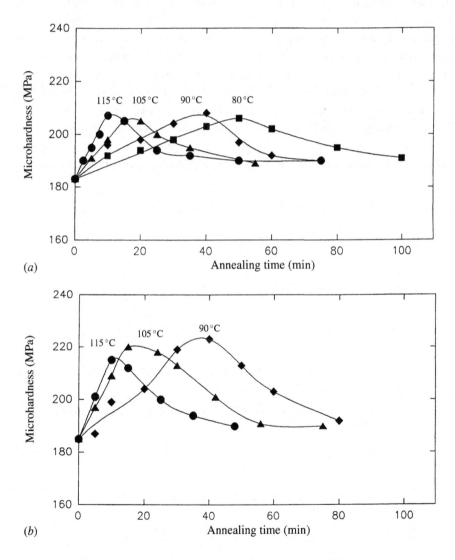

(*a*)

(*b*)

Figure 3.9. Microhardness as a function of annealing time for two differently aged PEN samples at different temperatures. Storage time: (*a*) 1 year, (*b*) 3 years. (After Rueda *et al.*, 1995.)

H_{max} seem to depend on the ageing of the initial samples and are independent of T_a. The final hardness values, H_f, obtained after longer annealing times are, however, larger than the H_0 values corresponding to the samples stored at room temperature. Finally, after 1–2 days storage of the 'dried' samples at room temperature in an ambient atmosphere, the low original H_0 values are recovered. Similar results are found for the other annealing temperatures (Rueda *et al.*, 1995).

What explanation can be given for the H_{max} values observed? In the first place, for short annealing times, the progressive loss of water, which acts as a plasticizer should contribute to the microhardening of the material. After removal of the water a maximum in H should be obtained (H_{max}) since for longer annealing times one might speculate that a decrease of H might be connected with a relaxation of the chain molecules in the 80–115 °C range caused by the onset of the libration motion of naphthalene moities (Troughton *et al.*, 1989; Ezquerra *et al.*, 1993).

After storage of the annealed sample for 1–2 days at room temperature, the film recovers the original degree of hydration and reaches its equilibrium state. Upon further annealing at high T_a, the same hardening cycle is reproduced.

A convenient property of the glassy matrix for determining extensive physical ageing is the excess enthalpy. Results from Rueda *et al.* (1995) indicate that the

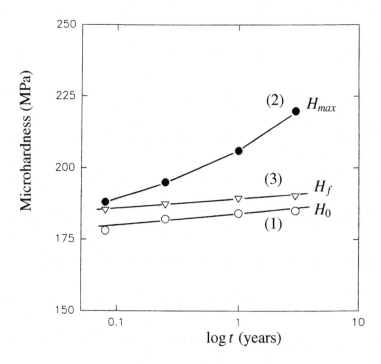

Figure 3.10. The dependence of microhardness upon storage time for PEN: (1) at room temperature, (2) subsequently annealed at $T_a = 105\,^\circ$C up to H_{max}; (3) subsequently annealed at $T_a = 105\,^\circ$C to the final value (after 60–100 min) (see text). (After Rueda *et al.*, 1995.)

excess enthalpy involved in the ageing process due to storage of the samples appears far below T_g and not at the T_g transition, as is often observed for other polymers with lower T_g values (Petrie, 1972; Cotwie & Ferguson, 1989). The presence of a sub-T_g endothermic peak has been explained using a distribution of relaxation times (Ramos *et al.*, 1984). However, after annealing near T_g, for short periods of time and/or at lower temperatures for prolonged periods of time, an increased endothermic peak is resolved just at the end of the glass transition interval as is observed for PET and other polymers. The excess enthalpy of the stored samples of an annealed sample supports the occurrence of physical ageing with storage in ambient atmosphere and that this process, as would be expected, is thermally enhanced (Struik, 1978). Additionally, for the very short periods of time that were investigated, an increase of the excess enthalpy with annealing time was observed.

3.3.2   Copolymers

The study of physical ageing of glassy polymers has been extended to copolymers. Exploring the well known approach of suppressing the crystallization ability of block copolymers by increasing the amount of non-crystallizable component, Giri *et al.* (1997) investigated films of poly(buthylene) terephthalate)-cyclo-aliphatic carbonate (PBT-PCc). Copolymers with increasing amount of PCc (up to 60 wt%) were prepared. DSC and wide-angle X-ray scattering (WAXS) measurements indicated that the quenched samples containing more than 20 wt% PCc were amorphous and crystallized only after appropriate annealing. The microhardness of two series of the glassy samples, both with variable composition, has been measured (i) immediately after solidification (fresh samples, H_a^{fresh}) and (ii) after storage for 6 months at room temperature (aged samples, H_a^{aged}). The aged samples clearly displayed larger microhardness values than the freshly prepared samples. What is more, for all amorphous copolymers, both H^{fresh} and H_a^{aged} linearly increased with the concentration of PCc units. The extrapolation of the two straight lines gave the H value of purely amorphous PBT. Two different values for H_a^{PBT} (54 MPa and 94 MPa) were obtained from the data for H_a^{fresh} and H_a^{aged}, respectively. In addition, the hardness additivity of the aged samples was preserved in contrast to the semicrystalline copolymers.

In conclusion, microhardness evaluation is a method capable of measuring the molecular reorganization taking place above and below T_g. On one hand it detects the contribution of chain mobility with increasing annealing time and temperature leading to a more compact structure (physical ageing). On the other hand, the hardness measures the influence of a thermal expansion.

3.4   Correlation between microhardness and the glass transition temperature

Yield mechanisms in glassy polymers have been reviewed by Crist (1993). The intrinsic stiffness of these rigid polymers below T_g leads to microhardness values which may even be 2–3 times larger than those obtained for typical flexible polymers. As we have seen at the beginning of the chapter, at the glass transition, there is an onset of liquid-like motions, involving longer segments. These motions of larger bundles, above T_g, require more free volume and lead to a fast decrease of microhardness with temperature.

It is known that T_g generally increases with increasing cohesive energy density (CED) according to the following equation (Gedde, 1995):

$$T_g = \frac{2\delta^2}{mR} + C_1 \tag{3.3}$$

where δ^2 is the CED, m is a parameter that describes the internal mobility of the groups in a single chain, R is the gas constant and C_1 is a constant. It is noteworthy that the CED provides an integrated measure of the strength of the secondary bonds in a compound; materials with strong secondary bonds show high CED values. Thus, eq. (3.3) accounts for the two most important factors determining the value of T_g: (1) the possibility for rotation around single bonds and (2) formation of secondary bonds with the surrounding atoms. In addition, it offers a link between the glass transition temperature and the microhardness of polymer glasses since the cohesion energy is the basic factor in determining also the microhardness (Martínez-Salazar et al., 1985).

By this common dependence of T_g and H on cohesion energy, one can explain the almost linear relationship between H and T_g.

The dependence of microhardness on the glass transition temperature for some commercial polymers is presented in Fig. 3.11. The polymers used for the plot in Fig. 3.11 are summarized in Table 3.2. Only non-crystallizable polymers have been selected. For the crystallizable ones, for which there are no reliable evidence for the presence of a completely amorphous phase, as is the case for example with polyamide 6 (PA6), the H_a value obtained by extrapolation to zero crystallinity is used. In this way we attempt to avoid the very strong effect on H of even small amounts of the crystalline phase.

One can see that a fairly good linear relationship between H and T_g exists in the T_g interval between 300 and 500 K, the regression coefficient being 0.96. For this reason one can write:

$$H = kT_g + C \tag{3.4}$$

where $C = -571$ MPa and $k = 1.97$ MPa/K. The value of C is physically meaningless. It only reflects the fact that polymers characterized by T_g values below room temperature do not have the commonly used microhardness, i.e. that measured at room temperature. In order to obtain physically meaningful values of H for these substances one has to decrease the temperature at which the measurements have to be carried out in such a way that $T_{measured} < T_g$ always. For this reason C has to be considered as a constant which helps to describe the microhardness behaviour of some complex systems in a better way as we shall see in Chapter 5. Thus, eq. (3.4) offers one the opportunity to characterize the contribution of the soft-segment phase to the overall microhardness of the system, as will be demonstrated in Chapter 5.

Finally, it is worth mentioning that the linear relationship between H and T_g has been derived only from one-phase systems (amorphous homo- and copolymers, Table 3.2). However, it can be applied to explain the micromechanical behaviour of multicomponent or multiphase systems containing at least one liquid-like component or phase (see Chapter 5). Another peculiarity of the polymers listed in Table 3.2 is that their main chains comprise only single C–C, C–O or C–N bonds

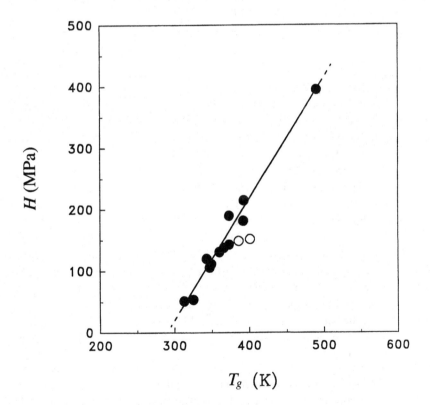

Figure 3.11. The relationship between the microhardness H and the glass transition temperature T_g for the amorphous polymers and copolymers listed in Table 3.2. The open circles correspond to samples 9 and 13 in Table 3.2 (see text). (From Fakirov *et al.*, 1999.)

(main chains containing aromatic rings such as for example polyimides disobey the linear relationship). This observation leads to the very preliminary conclusion that the linear relationship is valid mostly for amorphous polymers distinguished by T_g which is determined mainly by the rotation potential around single chemical bonds.

This assumption is supported by the behaviour of some of the data in Fig. 3.11. For example, the data points belonging to PC and the PET–PC copolymer (being the ones richest in PC in this series, open circles in Fig. 3.11) deviate to some extent from the linear relationship $H = f(T_g)$. If one disregards them the correlation coefficient of the straight line is 0.985 instead of 0.96 and the slope and coefficient C are still close to the above-cited ones.

The described deviation of the data for PC from the linear relationship (Fig. 3.11) can be explained by the fact that the chemical structure of these polymers differs significantly from that of polyolefines; instead of CH_2–CH_2 bonds, benzene rings

Table 3.2. *Glass transition temperature T_g and microhardness H of the polymers used for the plot $H = f(T_g)$ (Fig. 3.11).*

Sample No	Polymer	T_g (°C)	H (MPa)
1	Polyamide 6 (PA6)	40	52*
2	Poly(butylene terephthalate) (PBT)	52	54*
3	Poly(ethylene terephthalate) (PET)	70	120*
4	Random copolymer of PET and poly(1,4-cyclohexanediol terephthalate (PCHT) (PET/PCHT 64/36***)	74	106
5	Random copolymer of PET/PCHT 69/31***	76	111
6	Randomized copolymer of PET and bisphenol-A-polycarbonate (PET-PC 70/30**)	87	131
7	Randomized copolymer PET-PC 50/50**	93	138
8	Randomized copolymer of PET-PC 30/70**	100	143
9	Randomized copolymer of PET-PC 10/90**	113	149
10	Atactic polystyrene (a-PS)	100	190
11	Poly(ethylene naphthalane-2,6-dicarboxylate) (PEN)	119	182
12	Atactic poly(methyl methacrylate) (a-PMMA)	120	215
13	Bisphenol-a-polycarbonate (PC)	128	152
14	Dry gelatin	217	395

*   Value obtained by extrapolation to zero crystallinity

**   Composition in wt%. All the PET–PC copolymers are amorphous as shown by X-ray and DSC measurements and non-crystallizable as follows from their composition

*** Composition in mol%.

dominate in the main-chain structure. Obviously, for the same reason the data for polyarylates, which contain even more benzene rings than PC, display a behaviour far from the observed linear relationship and are not included in Table 3.2 and Fig. 3.11. One can expect that for polymers for which the T_g is not determined mostly by rotation around single bonds the relationship between H and T_g will be more complex.

Summarizing one can conclude that due to the empirical linear relationship between H and T_g in a rather broad range of T_g (−50 up to 250 °C) which covers most commonly used polymers of the polyolefin-type and also polyesters and polyamides, it is possible to calculate the microhardness value of any amorphous polymer provided its T_g is known ($H = 1.97T_g - 571$). Furthermore, one can account for the contribution of soft liquid-like components and/or phases (characterized by a negligibly small microhardness) to the microhardness of the entire system. As we shall see in Chapter 5 the plastic deformation mechanism of such systems is different from that when all the components and/or phases are solid, i.e. have T_g above room temperature.

3.5 Micromechanics of polymer glasses

3.5.1 Crazing in glassy polymers

Polymeric materials are used in many fields. In order to use them in structural components, it is necessary to recognize their mechanical strength. Two types of plastic deformation of amorphous polymers have been observed. One is shear flow followed by the formation of a localized deformed region, called a shear band. The other one is yielding caused by a number of fine fissures, called crazes. This type of deformation is significant in media such as organic liquids. Although to the naked eye a craze is similar to a crack, it is distinguished by its microscopic structure, which consists of many elongated fibrils and voids. Scission of the fibrils, however, turns the craze into a crack and leads the polymer to brittle fracture. Because crazes lead to plastic deformation and, finally, to polymer fracture, they exert a great influence on polymer strength. The phenomenon of crazing, therefore, has been extensively studied (Kramer, 1979; Kinloch & Young, 1983; Kausch, 1978; Kawagoe, 1996).

The microstructure and micromechanics of isolated crazes have drawn the attention of many investigators (see, for example, the review of Kramer & Berger (1990)), probably because of the similarity in shape between crazes and cracks. Kramer & Berger (1990) have observed craze microstructure by transmission electron microscopy (TEM) and analysed the stress–strain relationship in and around a craze on the basis of TEM images. They adopted a skilful technique to strain a copper grid coated with a thin film of a polymer to form crazes in the specimen chamber of an electron microscope. This technique has been adopted in other laboratories. The researchers revealed that there are two different mechanisms of craze thickening: the surface-drawing mechanism and the fibril creep mechanism.

In surface drawing, which is general for crazes grown in air, new fibrillar matter is drawn in from the surrounding unoriented polymer through a thin strain-softened layer called an active zone. In fibril creep, which is generally found in environmental crazes, the fibrils in a craze are further elongated. In air crazes, in particular, the fibrils are formed by disentanglement or scission of molecular chains in the active zone.

Wu (1990) has examined the relation of craze nucleation stress to chain entanglement density (the number of molecular entanglements per unit volume) for various polymers including crystalline polymers, and indicated a linear relationship between the logarithms of these two terms. In general, as entanglement density increases, so does the stress for craze nucleation, and the deformation mode varies from crazing to shear flow. According to Berger (1989) for amorphous polymers crazing takes place only for entanglement density smaller than 4×10^{25} chains m^{-3}, whereas shear flow occurs above 11×10^{25} chains m^{-3}. As the temperature is raised, the entanglement density that corresponds to the transition of the deformation mode becomes lower. Other mechanisms of chain scission participate more strongly in environmental crazing, in which environmental reagents plasticize an isolated area under stress concentration. The stresses in and around a single craze are derived from the displacement distribution measured by the TEM technique. Stress concentration is indicated at the craze tip. Stress analysis for a craze formed at a crack tip is also possible and reveals a great strain hardening in the craze fibrils.

Crazing mechanisms and criteria have long been studied (Kausch, 1978; Michler, 1992), but further work is still needed because of limited number of experimental results under multiaxial stress and considerable differences in crazing behaviour under different conditions (Kawagoe, 1996).

Analysing the stress and displacement of a craze formed at a crack tip is important for understanding brittle fracture. Recently, optical interference tests have been used in response to this problem. Craze opening displacement has been measured by this technique and the craze contour stress based on the Dugdale model of fracture mechanics has been obtained (Döll, 1983; Döll & Könczöl, 1990). These experiments showed the stress–strain relation in the craze zone at the crack tip and also demonstrated jumps in craze displacement under cyclic loading. These authors have also explained the mechanisms of retarded crack growth in fatigue processes. Schirrer (1990) has conducted similar measurements by using optical interferometry from a molecular viewpoint. He related the lifetime of craze fibrils to the craze contour stress calculated from the Dugdale–Döll–Könczöl model at different temperatures and revealed that the lifetime is related to the craze contour stress. Schirrer also found that β relaxation participates in the breaking of craze fibrils.

Direct measurements of craze contour displacement by SEM have also been developed. In addition to the above structural investigations, the craze fibril elongation ratio (Müller, 1963; Kambour, 1968, 1973) and the craze strength (Ishikawa *et al.*, 1996) of samples from the necking of elongated specimens were examined. In these studies the elongation of the fibrils was assumed to be equal to the elongation of the polymeric material in the necked region.

3.5.2 Micromechanics of crazes studied by ultramicrohardness

Michler *et al.* (1999) reported the first measurements of microindentations on craze fibrils in the micrometre range. It has been shown that the length of the craze zones detected by microhardness and the length of the crazes measured using interference optics are correlated. In addition, these authors calculated the elastic moduli of the crazed material from the elastic recovery curves of microindentations. In their experiments, commercially available samples of PS, poly(styrene–acrylonitril) copolymers (SAN), PVC and PMMA were used. The molecular weight, polydispersity and mechanical properties of these materials are listed in Table 3.3. The samples were actually miniaturized compact tensile (CT) samples ($10 \times 8 \times 4$ mm^3) with a crack tip craze in front of a sharp notch as shown in Fig. 3.12. The crazes were created during slow crack propagation under fatigue loading at room temperature. The microhardness was determined using a dynamic ultra-microhardness tester. The experiments were performed at room temperature. During microindentation the force was applied up to a maximum load of 10 mN at a constant loading rate of 1.4 mN s^{-1}. The penetration depth was less than 1.5 μm in all experiments. In all cases the maximum load was applied for 6 s (creep) and, then, the sample was unloaded at the same rate as in the typical loading–creep–unloading curve of Fig. 2.14. The microhardness value was calculated from the penetration depth and the applied load using eq. (1.2).

Table 3.3. *Molecular weight M_w, polydispersity M_w/M_n and mechanical properties of the bulk polymers investigated.*

Polymer	M_w (kg mol^{-1})	M_w/M_n	H_{bulk} (MPa)	E_{bulk} (GPa)
PMMA	2200	2.5	182 ± 2	4.8 ± 0.2
SAN	249	2.3	178 ± 2	4.7 ± 0.2
PS	476	4.2	164 ± 3	4.3 ± 0.1
PVC	181	1.7	145 ± 2	4.2 ± 0.2

The indentations were performed on a straight line along the direction of crack propagation (Fig. 3.12). The location of the indentation along the direction x was to referred with respect to the notch tip which was taken as the origin for the microhardness experiments. The diagonal of the indenter was located parallel to the crack/craze propagation direction. This is illustrated in Fig. 3.13. The first indentations were placed in the cracked region. The tip of the indenter dipped into the crack and the outer part of the indenter touched the bulk polymer. The following indentations were positioned in the crazed zone. In this case the tip of the indenter pushed the craze fibrils apart while the outer part pressed the undeformed polymer surrounding the craze. Finally the microhardness of the bulk polymer in front of the craze tip was also measured. The values of the bulk microhardness were derived from a mean of about 30 indentations.

The craze width at the area of indentation was measured by light microscopy and scanning electron microscopy.

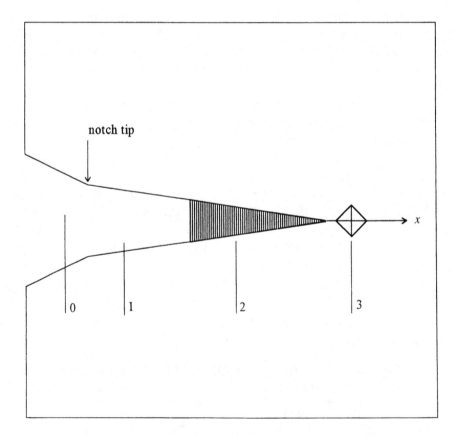

Figure 3.12. Schematic view on the CT sample illustrating the position of the notch (0), the crack (1) and the craze (2) regions; microhardness indentations were placed along the x straight line. (From Michler *et al.*, 1999.)

To derive the microhardness of the crazed region from the experimental indentation value a two-phase system was assumed. The experimental microhardness (H_{exp}) was assumed to be equal to the sum of the microhardnesses of the bulk polymer (H_{bulk}) and the craze (H_{craze}) weighted in their volume fractions at one indentation:

$$H_{exp} = (V_{bulk} H_{bulk} + V_{craze} H_{craze}) / V_{exp} \qquad (3.5)$$

with

$$V_{exp} = V_{bulk} + V_{craze} \qquad (3.6)$$

where V_{exp} is the volume of the entire indentation, V_{bulk} is the volume fraction of the bulk polymer and V_{craze} is the volume fraction of the craze zone.

The fibrils within the craze occupy only a certain portion of the craze volume, the fibril volume content v_f. To calculate the microhardness of the fibrils (H_{fibril}) the expression

$$H_{fibril} = v_f H_{craze} \qquad (3.7)$$

was used.

The elastic modulus (E) was calculated directly from the experimental loading–creep–unloading curve (as in Fig. 2.14) according to the procedure of Doerner & Nix (1986).

A procedure similar to the one described above was used to calculate the elastic moduli of the craze fibrils (E_{fibril}). The elastic modulus calculated directly from the experimental data (E_{exp}) was estimated to be the sum of the modulus of the bulk

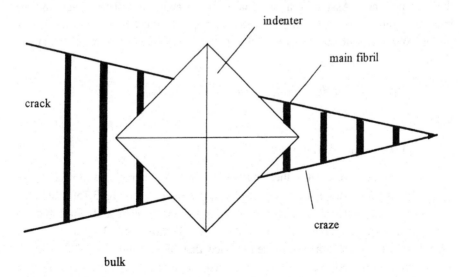

Figure 3.13. Schematics of the indentation on a craze (top view).

polymer (E_{bulk}) and the modulus of the craze, weighted in their volume fractions. The fibril volume content in the craze was taken into account when calculating the modulus of the craze fibrils (E_{fibril}). The elastic modulus was then given by eq. (3.8).

$$E_{exp} = (V_{bulk}E_{bulk} + V_{craze}E_{fibril}/v_f)/V_{exp} \tag{3.8}$$

The contour of the loaded and unloaded craze was investigated by interference optics (Döll & Könczöl, 1990; Könczöl et al., 1990). Additionally the structure of the crazes was investigated by high-voltage TEM. Sections of about 1 μm were microtomed for TEM observations. Structural inspection of the craze zone was done by TEM. The fibrillar structure of a craze zone in SAN is shown in Fig. 3.14. The microstructure of this craze created in a CT sample is similar to the structure investigated *in situ* in strained semithin sections (Michler, 1990).

3.5.3   Microhardness and elastic moduli of glassy microfibrils

Figure 3.15 illustrates the microhardness experimental data from various indentations on a PS specimen. The horizontal solid line represents the microhardness average value of the bulk polymer and the dotted lines represent the zone of scattering data. The experimental results show three distinct regions of behaviour: (*a*) crack zone, (*b*) craze region and (*c*) bulk.

The H values measured along the crack zone are lower than the microhardness of the bulk. In this case the indentations were made close to the notch tip. The indenter tip was placed into the crack and the outer part of the indenter pressed on the bulk polymer. Assuming a two-phase system consisting of the crack and the bulk material, a lower H value than that of the bulk must be measured because the microhardness of the cracked portion (H_{crack}) is equal to zero (eqs. (3.9) and (3.10)).

$$H_{exp} = (V_{bulk}H_{bulk} + V_{crack}H_{crack})/V_{exp} < H_{bulk} \tag{3.9}$$

with

$$V_{exp} = V_{bulk} + V_{crack} \quad \text{and } H_{crack} = 0 \tag{3.10}$$

It has been observed that in these particular samples the crack is followed by the craze. The microhardness increases rapidly in this region (see Fig. 3.15). Table 3.3 shows the bulk microhardness values of the polymers investigated. In Table 3.4 the H values of the craze fibrils are listed. The data of Table 3.4 show that the microhardness of the fibrils is at least twice that of the bulk polymer. This is consistent with the concept of highly oriented polymer chains within craze fibrils (Bin Ahmad & Ashby, 1988). It is noteworthy that the microhardness of craze fibrils in amorphous polymers is of the same order of magnitude as the microhardness of microfibres of semicrystalline PET (Krumova *et al.*, 1998). It is also to be noted that the craze strength, as a parameter capable of characterizing craze stability, is comparable to the fibril hardnesses in amorphous polymers (Ishikawa *et al.*, 1996).

If we consider the PMMA, SAN and PS samples, then, it is seen that the quotients H_{fibril}/H_{bulk} decrease with decreasing microhardness of the bulk polymer (see

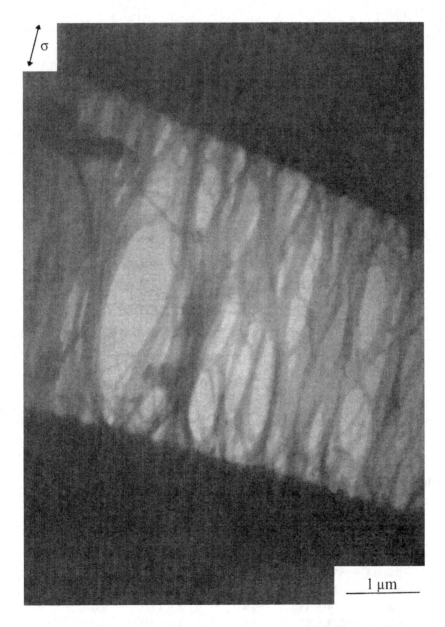

Figure 3.14. TEM micrograph of the fibrillar craze structure in SAN. (From Michler *et al.*, 1999.)

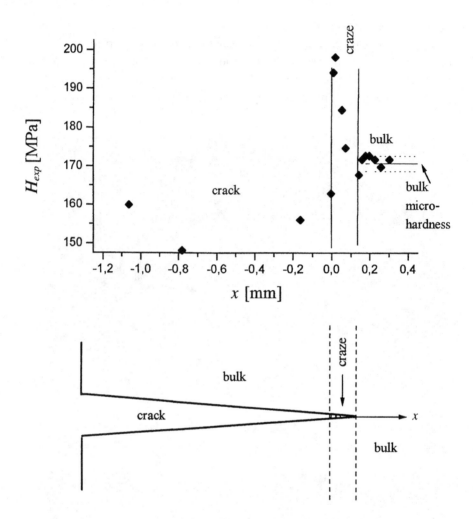

Figure 3.15. Microhardness values H_{exp} measured along the crack propagation direction x, in PS. Cracked, crazed and bulk zones are shown. (From Michler *et al.*, 1999.)

Table 3.4. *Microhardness H and elastic moduli E of the craze fibrils as well as the ratio of these values to the bulk properties of diferent polymers.*

Polymer	H_{fibril} (MPa)	H_{fibril}/H_{bulk}	E_{fibril} (GPa)	E_{fibril}/E_{bulk}
PMMA	537 ± 16	3.0	13.4 ± 1.1	2.8
SAN	510 ± 20	2.9	12.2 ± 1.4	2.6
PS	414 ± 33	2.4	11.2 ± 1.6	2.6
PVC	698 ± 82	4.8	11.4 ± 1.5	2.7

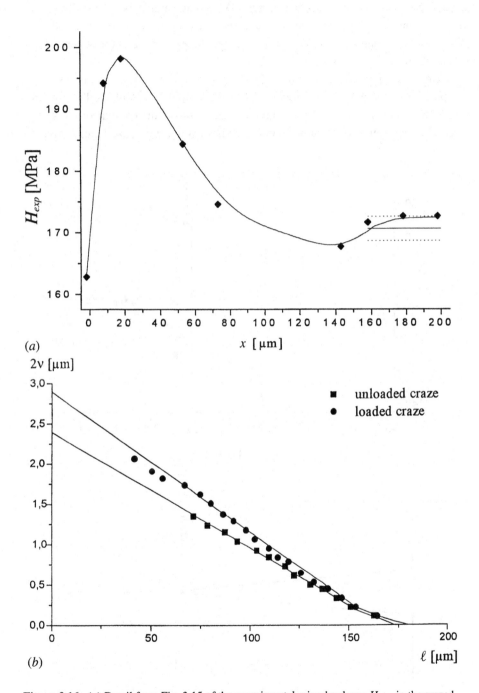

Figure 3.16. (a) Detail from Fig. 3.15 of the experimental microhardness H_{exp} in the crazed zone. (b) Craze opening ($2v$) vs length (ℓ) of the same craze shown in (a) measured by interference optics. (From Michler *et al.*, 1999.)

Table 3.4). This is because of the correlation between the mechanical properties of each polymer and those of the crazes. A similar parallel behaviour between the microhardness values of the bulk and the craze fibrils was not observed for PVC. This discrepancy is probably due to the presence of additives (polymerization initiators, stabilizers, etc.).

The craze length derived from microhardness indentation measurements has been compared with the length measured by interference optics. An enlarged section of Fig. 3.15 is shown in Fig. 3.16(*a*). Figure 3.16(*b*) shows the contour of the same craze measured by optical interferometry. It is seen that the craze length derived

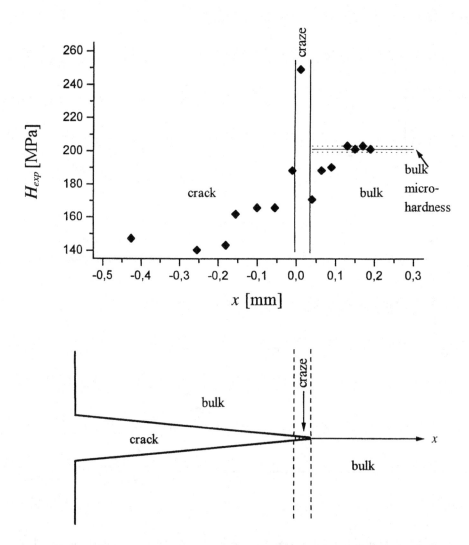

Figure 3.17. Microhardness H_{exp} derived from different indentations in PMMA along the direction of craze propagation x. (From Michler *et al.*, 1999.)

from the microhardness experiments is slightly smaller than the interferometrically measured length. This is due to the lower resolution of the microhardness measurements close to the craze tip. In any case the agreement between the values is acceptable. Another interesting finding should be pointed out: the experimental microhardness values of indentations in the area near the craze tip are smaller than the bulk microhardness (see Fig. 3.17). From electron microscopy it is known that the craze fibrils are broken in the vicinity of the sample surface (Michler *et al.*, 1999, unpublished). The explanation proposed for these low microhardness values is that near to the craze tip the indenter does not touch the craze fibrils because of the small indentation depth. Therefore the superposition of normal and lateral forces at the indenter tip can give smaller experimental microhardness values. Nevertheless, the microhardness measurement is shown to be successful in the detection of very short crazes. Figure 3.17 clearly illustrates indeed the location of a craze in PMMA shorter than 30 μm. Figure 3.18 shows the contour of the same craze detected by interference optics.

For all polymers investigated it has been shown that the fibril moduli is, at least, twice than the value of the bulk. A proportionality between the elastic moduli of the bulk and of the craze fibrils is not obvious. This may be due to the lower accuracy in the calculation of this parameter.

Figure 3.19 shows the dependence of the elastic modulus of craze fibrils E_{fibril} of PMMA, SAN and PS on the entanglement density (ν_e). It is known that the increasing density of entanglement in a polymer induces an increase of the bulk

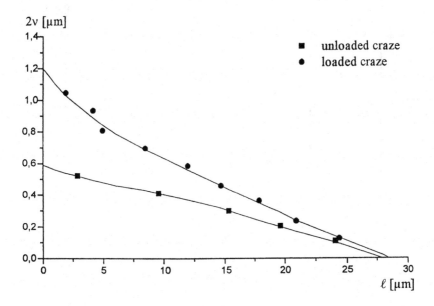

Figure 3.18. Craze opening (2ν) *vs* length (ℓ) for the same craze in PMMA as in Fig. 3.17 measured by interference optics. (From Michler *et al.*, 1999.)

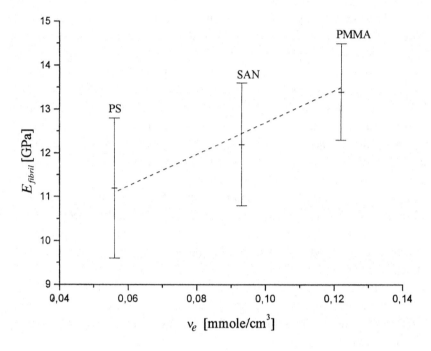

Figure 3.19. Elastic modulus of the craze fibrils E_{fibril} *vs* the entanglement density ν_e for PS, SAN and PMMA. (From Michler *et al.*, 1999.)

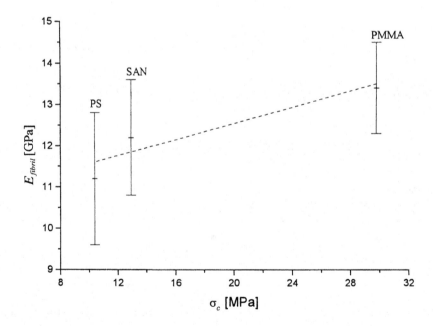

Figure 3.20. Elastic modulus of the craze fibrils E_{fibril} *vs* the crazing stress σ_c for PS, SAN and PMMA. (From Michler *et al.*, 1999.)

elastic modulus (Aharoni, 1989; Elias, 1993). In agreement with this concept the results of Fig. 3.19 show that the modulus of the craze fibrils also increases with entanglement density. Most interesting, however, is the result that the elastic modulus of the craze fibrils is proportional to the critical stress required to create the crazes (crazing stress, σ_c) as can be concluded from Fig. 3.20. This finding is, in turn, in accordance with the known proportionality between crazing stress and the molecular entanglement weight (Wu, 1990). A remaining question is the dependence of both the fibril microhardness and the fibril elastic modulus on the loading history.

In summary, it can be concluded that using nanoindentation hardness measurements on the crack tip, down to penetration depths of 0.8 μm, it is possible to detect very small craze zones in glassy polymers. The microhardnesses for all investigated samples can be divided into three regions: (1) the cracked region, (2) the crazed zone and (3) the bulk material. It was also found that the microhardness of the crazed material is larger than the microhardness of the bulk polymer due to the orientation of the polymer chains within the craze fibrils.

The elastic modulus in the craze is also larger than that of the bulk. These results underline the role of crazes as the precursors of cracks, conferring to the polymer more resistance against crack propagation. Furthermore, it has been shown that the elastic modulus of craze fibrils (PMMA, SAN, PS) is proportional to both the crazing stress and the entanglement density. How the microhardness of craze fibrils in PVC relates to the matrix properties is not yet understood. Additives strongly affect the experimental data giving rise to a softening of the bulk and of the fibrils. It is also possible to estimate the craze length from microhardness indentation measurements. However the result is less accurate than that obtained by other techniques.

3.6   References

Aharoni, S. (1989) *Macromolecules* **19**, 426.

Ania, F., Kilian, H.G. & Baltá Calleja, F.J. (1986) *J. Mater. Sci. Lett.* **5**, 1183

Ania, F., Martínez-Salazar, J. & Baltá Calleja, F.J. (1989) *J. Mater. Sci.* **24**, 2934.

Baltá Calleja, F.J. (1985) *Adv. Polym. Sci.* **66**, 117.

Baltá Calleja, F.J., Fakirov, S., Roslaniec, Z., Krumova, M., Ezquerra, T.A. & Rueda, D.R. (1998) *J. Macromol. Sci. Phys.* **B37**, 219.

Berger, L.L. (1989) *Macromolecules* **22**, 3162.

Bin Ahmad, Z. & Ashby, M.F. (1988) *J. Mater. Sci.* **23**, 2037.

Bowden, P.B. & Raha, S. (1973) *Phil. Mag.* **31**, 149.

Brandrup, Y. & Immergut, E.H. (1989) *Polymer Handbook*, third edition, John Wiley & Sons, New York.

Briscoe, B.J., Sebastian, K.S. & Adams, M.J. (1994) *J. Phys. D: Appl. Phys.* **27**, 1156.

Cotwie, J.M.G. & Ferguson, R. (1989) *Macromolecules* **22**, 2307.

Crist, B. (1993) in *Materials Science and Technology*, Vol. 12 (Cahn, R.V., Haasen, P. & Kramer, E.J., eds.) VCH, Weinheim, p. 437.

Doerner, M.F. & Nix, W.D. (1986) *J. Mater. Res.* **1**, 601.

Döll, W. (1983) *Adv. Polym. Sci.* **52/53**, 105.

Döll, W. & Könczöl, L. (1990) *Adv. Polym. Sci.* **91/92**, 136.

Elias, H.-G. (1993) *An Introduction to Plastics*, VCH, Weinheim, p. 149.

Ezquerra, T.A., Baltá Calleja, F.J. & Zachmann, H.G. (1993) *Acta Polymerica* **44**, 18.

Fakirov, S. (1978) *Colloid & Polymer Sci.* **256**, 115

Fakirov, S., Apostolov, A.A., Boesecke, P. & Zachmann, H.G. (1990) *J. Macromol. Sci. Phys.* **B29**, 379.

Fakirov, S. Baltá Calleja, F.J. & Krumova, M. (1999) *J. Polym. Sci. Polym. Phys.* **37**, 1413.

Fakirov, S., Cagiao, E., Baltá Calleja, F.J., Sapundjieva, D. & Vassileva, E. (1999) *Int. J. Polym. Mater.* **43**, 195.

Fakirov, S., Fakirov, C., Fischer, E.W. & Stamm, M. (1991) *Polymer* **32**, 1173.

Fakirov, S. & Gogeva, T. (1990) *Makromol. Chem.* **191** 615.

Gedde, U.W. (1995) *Polymer Physics*, Chapman & Hall, London.

Giri, L., Roslaniec, Z., Ezquerra, T.A. & Baltá Calleja, F.J. (1997) *J. Macromol. Sci. Phys.* **B36**, 335.

Ion, R.H., Pollock, H.M. & Roques-Carmes, C. (1990) *J. Mater. Sci.* **25**, 1444.

Ishikawa, M., Ushui, K., Kondo, Y., Hatada, K. & Gima, S. (1996) *Polymer* **37**, 5375.

Kambour, R.P. (1968) *Appl. Polym. Symp.* **7**, 215.

Kambour, R.P. (1973) *J. Polym. Sci. Macromol. Rev.* **7**, 1.

Kausch, H.H. (1978) *Polymer Fracture*, Springer-Verlag, Berlin, Chapter 9.

Kawagoe, M. (1996) in *Polymeric Materials Encyclopedia*, Vol. 4 (Salamone, J.C. ed.) CRC Press, Boca Raton, New York, p. 2807.

Kinloch, A.J. & Young, R.F. (1983) *Fracture Behaviour of Polymer*, Applied Science, London, Chapter 5.

Kramer, E.J. (1979) *Developments in Polymer Fracture* (Andrews, E.H., ed.) Applied Science, London, Chapter 3.

Kramer, E.J. & Berger, L.L. (1990) *Adv. Polym. Sci.* **91/92**, 1.

Krumova, M., Baltá Calleja, F.J., Fakirov, S. & Evstatiev, M. (1998) *J. Mater. Sci.* **33**, 2857.

Legge, N.R., Holder, G. & Schroeder, H.E. (eds.) (1987) *Thermoplastic Elastomers. A Comprehensive Review*, Hanser Publishers, Munich.

Mark, J.E. (1996) *Physical Properties of Polymers Handbook*, AIP Press, Woodbury, New York.

Martínez-Salazar, García Peña, J. & Baltá Calleja, F.J. (1985) *Polym. Commun.* **26**, 57.

Michler, G.H. (1990) *J. Mater. Sci.* **25**, 2321.

Michler, G.H. (1992) *Kunststoff-Mikromechanik*, Carl Hanser, Munich, p. 88.

Michler, G.H., Ensslen, J., Baltá Calleja, F.J., Könczöl, L. & Döll, W. (1999) *Phil. Mag.* **A79**, 167.

Müller, F.H. (1963) *Materialprüfung* **5**, 336.

Petrie, S.E.B. (1972) *J. Polym. Sci.* **A2**(10), 1225.

Pharr, G.M. & Oliver, W.C. (1992) *MRS Bullet.* **07/28**.

Ramos, A.R., Hutchinson, J.M. & Kovacs, A. (1984) *J. Polym. Sci. Polym. Phys. Ed.* **22**, 1655.

Roe, R.-J. (1990) in *Concise Encyclopedia of Polymer Science and Engineering* (Kroschwitz, J.I. ed.) Wiley Interscience, New York,. p. 433.

Rose, P.I. (1987) in *Encyclopedia of Polymer Science and Engineering*, Vol. 7, second edition (Mark, H.F., Bikales, N.M, Overberger, C.G. & Menges, G. eds.) J. Wiley & Sons, New York, p. 488.

Rueda, D.R., Martínez Salazar, J. & Baltá Calleja, F.J. (1985) *J. Mater. Sci.* **20**, 834.

Rueda, D.R., Varkalis, A., Viksne, A., Baltá Calleja, F.J. & Zachmann, H.G. (1995) *J. Polym. Sci. Polym. Phys. Ed.* **33**, 1653.

Rueda, D.R., Viksne, A., Malers, L., Baltá Calleja, F.J. & Zachmann, H.G. (1994) *Macromol. Chem. Phys.* **195**, 3869.

Schirrer. R. (1990) *Adv. Polym. Sci.* **91/92**, 215.

Schroeder, H. & Cela, R. (1988) in *Encyclopedia of Polymer Science and Engineering*, Vol 12, John Wiley & Sons, New York.

Simov, D., Fakirov, S. & Mikhailov, M. (1970) *Colloid & Polym. Sci.* **238**, 521

Stribeck, N., Sapundjieva, D., Denchev, Z., Apostolov, A.A., Zachmann, H.G. & Fakirov, S. (1997) *Macromolecules* **30**, 1329.

Struik, L.C.E. (1978) *Aging of Amorphous Polymers and Other Materials*, Elsevier, Amsterdam.

Struik, L.C.E. (1983) *Mechanical Failure of Plastics* (Brostow, W. & Corneliussen, R.D. eds.) Hanser, Munich, Chapter 15.

Struik, L.C.E. (1987) *Polymer* **28**, 1521.

Struik, L.C.E. (1990) in *Concise Encyclopedia of Polymer Science and Engineering* (Kroschwitz, J.I. ed.) Wiley Interscience, New York, p. 36.

Troughton, M.J., Davies, G.R. & Ward, I.M. (1989) *Polymer* **30**, 58.

Vassileva, E., Baltá Calleja, F.J., Cagiao, M.E. & Fakirov, S. (1998) *Macromol. Rapid Commun.* **19**, 451.

Voigt-Martin, I. & Wendorff, J. (1990) in *Concise Encyclopedia of Polymer Science and Engineering* (Kroschwitz, J.I. ed.) Wiley Interscience, New York, p. 50.

Wu, S. (1990) *Polym. Eng. Sci.* **30**, 753.

Chapter 4

Microhardness of crystalline polymers

4.1 Introduction to crystalline polymers

4.1.1 Polymer crystallization

Under special conditions, amorphous polymers may partially crystallize and become semicrystalline materials. The coexistence of crystalline and amorphous regions is observed in most synthetic polymers that are not obtained by polymerization in the solid state of crystalline monomers or cyclic oligomers. The amorphous state is often called the liquid state, although it may differ quite appreciably from a normal low-molecular-weight liquid. The dominance of the long, randomly coiled, linear macromolecules with the enormous anisotropy of their force field and the presence of crystals strongly affect the behaviour of the chains in the amorphous component. Virtually all amorphous chain sections have one end (cilia) or both ends (folds, tie molecules connecting two different crystals) fixed in the crystal lattice. Hence, the properties of the polymer chains in the surface layer between the crystals and the amorphous layers are substantially different from those in truly amorphous materials without any crystals. Chains of flexible polymers crystallize if they are sufficiently regular and not too hampered in the liquid state by entanglements. Thus a regular, polydisperse, medium-molecular-weight ($<100\,000$), highly regular PE, without side chains, crystallizes very easily.

   In a phase transformation such as crystallization, the two basic processes are the initiation or nucleation process by which a new phase is initiated within a parent phase, and the subsequent growth of the new phase at the expense of the previous one (Gedde, 1995). The former may occur homogeneously, by statistical fluctuations in the parent phase, or by formation of nuclei catalysed by heterogeneity or impurities present in the melt. Crystallization from the molten state proceeds at a finite rate at temperatures well below the melting temperature, under conditions far from equilibrium. The isothermal rate at which crystallinity develops in a pure homopolymer follows a universal pattern. One observes a sigmoidal rapid increase of the degree of crystallinity w_c followed by a very slow increase (Fig. 4.1). The comparatively rapid initial increase is called primary crystallization while the following much slower process is called secondary crystallization (Zachmann & Stuart, 1960). In the polarizing microscope, one very often observes that during primary crystallization spherulites grow until they impinge on each other and finally fill the sample completely (Fig. 4.2). During primary crystallization, to a first approximation one can assume that the degree of crystallinity within the spherulites is constant. The sigmoidal form of the curve in Fig. 4.1 can then be explained in terms of Avrami theory (Avrami, 1939; Evans, 1945). In this theory one takes into account the effect of statistically distributed morphological units, like spherulites with the assumption that on impingement growth ceases. One then obtains:

$$w_c(t) = 1 - \exp(-Gt^n) \tag{4.1}$$

where $w_c(t)$ is the degree of crystallinity at time t, G is the Avrami constant, which is proportional to the number of nucleating sites and n is the Avrami exponent, which determines the type of crystallization that is taking place. Because of the simplifications introduced in the Avrami equation, details of the growth geometry and type of nucleation cannot be elucidated from the isotherms by specifying the value of the exponent n. However, analysis of the overall crystallization data for

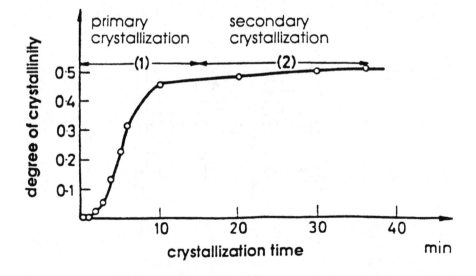

Figure 4.1. Typical plot of the degree of crystallinity as a function of time during the crystallization of polymers. (From Zachmann & Stuart, 1960.)

various polymers shows good adherence to this equation for the initial part of the crystallization process. The value of n can be obtained from the slope of the double-logarithmic plot of $\log w_c$ against $\log t$. Integral values of $n = 4$ and $n = 3$ have been found in many polymers (Wunderlich, 1976). However, in other cases, fractional exponents from data fitting have been reported (Connor *et al.*, 1997).

4.1.2   Morphology

The unoriented crystalline polymer obtained by solidification of the undercooled isotropic melt at rest is characterized by the dense stacking of crystalline lamellae about 10–20 nm thick with transverse dimensions of 0.1–1 μm (Bassett, 1981). The lamellae present uniform thicknesses as evidenced by electron microscopy

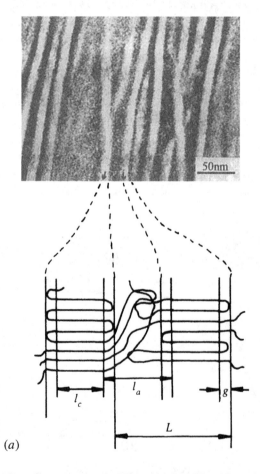

(a)

Figure 4.2. (*a*) (top) Lamellar structure of a PE section crystallized from the melt. (bottom) Schematics of parameters determined on micrographs. (From Baltá Calleja *et al.*, 1977.) (*b*) Photomicrograph of banded poly(trimethylene glutarate) spherulites crystallized from the melt. (From Keller, 1959.)

(Fig. 4.2(a)). The lamellae are, in turn arranged in spherulites (several tens of micrometres in diameter growing radially by small-angle non-crystallographic branching in all directions) (Fig. 4.2(b)).

Crystallization increases density and reduces the free volume. In moulds, the crystallization starts at the surfaces where the melt is cooler. Crystallization may be normal although columnar crystallization occurs preferentially adjacent to the cold surface. The interior crystallizes subsequently with an appreciable reduction in volume. As a result there is a tensile stress in the centre of the moulded specimen and a compressive stress on the surface. These stress fields influence the mechanical properties.

Under high pressure polymers, especially PE, may crystallize from the melt in the form of so called extended-chain crystals whose average lamella thickness ℓ_c of 150 nm is much larger than the usual ℓ_c of 20 nm and more comparable to the extended macromolecule length (Bassett, 1981). Such crystals are not usually made of fully extended chains, but contain folds that are less numerous the thicker the crystals and the lower the molecular weight.

An interesting feature of crystalline polymers is their composite structure. In semicrystalline solids, the polymer chains in the amorphous part are intimately connected with the crystals and cannot act independently (Fig. 4.2(a) bottom). The crystalline and the amorphous components depend on each other in a way that is different from the situation in non-polymeric composites. The chain mobility is limited by the crystals or other solids to which the chains are anchored. This causes a

(b)

Figure 4.2. Continued.

volume change during uniaxial extension or compression in total disagreement with the infinite rubber model in which the volume remains constant with deformation. Amorphous molecules flow in from the sides in order to achieve constancy. In a polymeric semicrystalline solid the amorphous chains cannot move as far in the long and thin amorphous layers between the crystals. Consequently, the deformation of the semicrystalline solid polymer must include in the elastic modulus of the amorphous component both moduli of rubber: the large bulk modulus K that takes into account the volume changes and the small shear modulus G that does not.

In addition, most semicrystalline polymers, particularly those produced commercially, are partially oriented; i.e. their chains have an overall alignment that may impart to the bulk polymer certain advantageous properties, e.g. increased mechanical strength or dielectric polarizability. Molecular orientation, whether arising from crystallization under stress or deformation of a solidified polymer, or in naturally occurring oriented crystalline polymers such as cellulose or keratin, is always associated with an orientational morphology.

The crystal size depends on the crystallization temperature and the molecular weight. Consequently, a wide range of sizes may be developed. The crystal thickness can be determined from small-angle X-ray scattering (SAXS) or from analysis of the Raman low-frequency longitudinal mode. The latter method also gives the crystallite size distribution. Thicker crystallites are obtained when crystallization takes place under isothermal conditions with a complex thickening and a variation in size distribution. Structural differences in the lamellae-like crystallites may also be due to the molecular weight and crystallization temperature.

Single-crystal mats are obtained by precipitation and subsequent filtration on a flat surface of the polymer from dilute solution (Bassett, 1981). Individual lamellae are formed with few connections from highly dilute solutions. In such a case, the

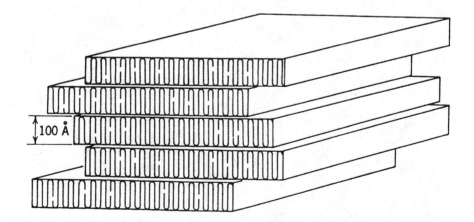

Figure 4.3. Schematics of an oriented single-crystal mat of a polymer as grown from solution. (From Ingram & Peterlin, 1968.)

mat is extremely brittle with little lateral cohesion (Fig. 4.3). The relevant parameter of the crystalline polymer is the lamellar periodicity (long period) $L = \ell_c + \ell_a$ of the lamellae stacking with interlamellar amorphous thickness ℓ_a (Fig. 4.2(a)). Crystal thickness ℓ_c depends primarily on undercooling $\Delta T = T_m^0 - T_c$, and therefore on crystallization temperature T_c. Here T_m^0 is the melting point or the solution point of an infinite crystal in a solvent at 101.3 kPa (1 atm). Any temperature changes during crystal growth are reflected in the lamellae thickness ℓ_c or long period L. Abrupt temperature changes produce small steps.

4.1.3 Polymorphism

A peculiarity of crystalline polymers, in contrast to the amorphous ones, is their ability to exist in different crystal modifications and thus to undergo polymorphic transitions (crystalline phase transitions). The various types of crystal modifications can be obtained by changing the crystallization conditions, such as temperature, cooling rate, etc.

These crystal modifications differ in their molecular and crystal structures as well as in their physical properties. Many types of crystalline modifications are reported, including a stable orthorhombic phase and metastable monoclinic phase for PE; α, β and γ forms for isotactic polypropylene (*i*-PP); trigonal and orthorhombic phases for polyoxymethylene; α and γ forms for Nylon 6; and others. Poly(vinylidene fluoride) (PVF), for example, appears in at least four types of crystalline modification (Lovinger, 1985; Dunn & Carr, 1989).

The most common case of the polymorphic transition is the thermally-induced crystalline phase transition. For example, PTFE undergoes a reversible two-step phase transition at 19 and 30 °C. The molecular conformation changes from a (13/6) helix to a (15/7) helix above 19 °C, whereas the unit cell changes from triclinic to hexagonal. In the transition at 30 °C, the conformation is disordered, although the hexagonal packing is maintained. In the crystal regions of *i*-PP and isotactic polystyrene (*i*-PS), the disorder-to-order phase transition occurs irreversibly by annealing. In cold-drawn *i*-PP, the (3/1) helices are packed upward and downward in a statistically disordered fashion (space group C2/c). As the annealing temperature increases, this packing disorder is gradually converted to the ordered phase where the upward and downward helices are packed in a regular fashion.

Vinylidene fluoride–trifluoroethylene (VF$_2$–F$_3$E) copolymers exhibit a ferroelectric–paraelectric phase transition, the first such case found for a synthetic polymer. In this transition, the electric polarization and piezoelectric constant of the film disappear above the Curie point (T_{Curie}). The temperature dependence of the dielectric constant, ϵ, obeys the so called Curie–Weiss law:

$$\epsilon = \epsilon_0 + \frac{C}{T - T_{Curie}} \tag{4.2}$$

where C is a constant and T is the temperature. Figure 4.4 shows the crystal structure of the ferroelectric (low-temperature) and paraelectric (high-temperature) phases. The ferroelectric phase has a polar structure consisting of planar-zigzag chains with the CF_2 dipoles arranged parallel to the b axis of the unit cell, while in the paraelectric phase the molecular chains, which are constructed by a statistical combination of TG, $T\bar{G}$, and TT rotational isomers and so contract from the extended planar conformation, rotate violently around the molecular axis and

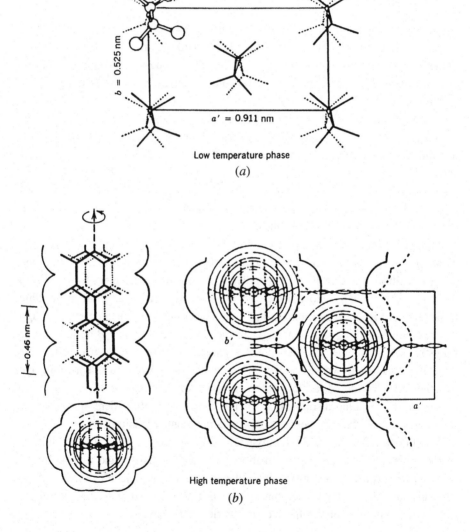

Figure 4.4. Crystal structure of (a) the ferroelectric and (b) the paraelectric phases of VF$_2$–F$_3$E copolymer. (From Tashiro & Tadokoro, 1987.)

are packed in a non-polar unit cell of hexagonal symmetry. The large molecular conformational change which occurs in this phase transition is accompanied by a rearrangement of the CF_2 dipoles in the unit cell, resulting in the remarkable change of the mechanical and electrical properties in the ferroelectric phase transition. This phase transitional behaviour changes with the VF_2 content of the copolymers (Tashiro & Tadokoro, 1987).

Polymorphic transitions are known to be induced by an external field such as tensile stress, hydrostatic pressure or shearing stress, high electric field and others. Crystalline phase transitions under tensile stress are discussed in Chapter 6 in relation with the observed changes in microhardness.

4.2   Microhardness of lamellar structures

4.2.1   Deformation mechanism

In contrast to low-molecular-weight solids, built up by an agglomeration of randomly oriented crystallites, the polycrystalline polymeric solid shows, as pointed out in the preceding section, a distinct lamellar morphology of crystalline lamellae intercalated by so called 'amorphous', less ordered, regions (Kanig, 1975; Grub & Keller, 1980; Bassett, 1981). The flat shape of lamellae favours parallel packing into crystal stacks usually producing a local ordering and orientation for a few lamellar thicknesses (Fig. 4.2(a)). The lamellae themselves have a mosaic block superstructure (Hosemann $et\ al.$, 1966). The concentration of lattice defects at the boundary between adjacent blocks reduces the lateral cohesion of stacked lamellar in the solid (Peterlin, 1971, 1973). Nevertheless, the connection of crystalline lamellae through a network of tie molecules and entanglements significantly contributes to the mechanical strength of the polymeric solid (Ward, 1982).

Figure 4.5(a) shows the three principal regions of a semicrystalline polymer; namely the crystalline region with a three-dimensional ordered structure, the interfacial region and the interzonal or amorphous regions, which consist of chain units in non-ordered conformations that connect crystallites.

When the polymeric material is compressed the local deformation beneath the indenter will consist of a complex combination of effects. The specific mechanism prevailing will depend on the strain field depth round the indenter and on the morphology of the polymer. According to the various mechanisms of the plastic deformation for semicrystalline polymers (Peterlin, 1971), the following effects may be anticipated:

(a) Lamellar fracture at the block (D_c) boundaries, while chains bridging the fractured blocks become partially unfolded and total cooperative block destruction at very large strains ($>20\%$), i.e. at depths $\sim 0.2h$, where h is the penetration of the indenter into the sample.

(b) Interlamellar sliding and separation involving shearing and compressional deformation of the amorphous layers and partial destruction of some blocks at slightly smaller strains (10–20%), at penetrations smaller than 0.5h.

(c) Phase transformation or twinning within the lamellar and elastic bending of crystals involving small strains (<10%), i.e. at depths equal to or larger than 0.5h.

When the magnitude of the stress reaches the yield point a macroscopic local plastic deformation is produced. The material beneath the indenter becomes permanently displaced and a microimpression arises. When the applied stress field

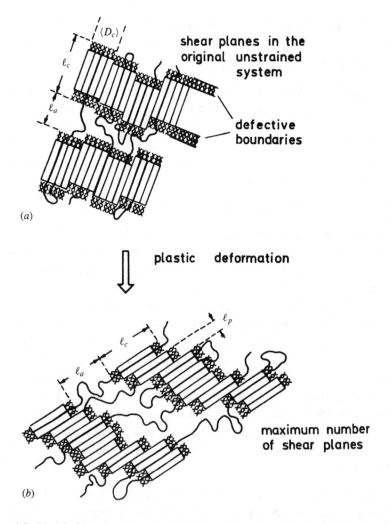

Figure 4.5. Model of lamellar structure: (a) before deformation, (b) after deformation under the stress field of the indenter. (After Baltá Calleja & Kilian, 1985.)

is removed, the molecules occluded in the amorphous layer and the network of molecular ties and/or entangled molecules acting as cross-links between adjacent more or less fractured crystals tend slowly to relax back. This effect contributes to the long-delayed elastic recovery of the material. The critical molecular weight for entanglement formation leading to a coherent molecular network structure is 30 000 (Kausch & Jud, 1982; van Krevelen, 1976). Below this figure no long-delayed recovery of the unloaded material is expected. Figure 4.5(b) schematically illustrates the suggested model of local plastic deformation of the compressed lamellae beneath the indenter to account for the above effects (Baltá Calleja & Kilian, 1985).

From a mechanical point of view the polymer may be regarded as a composite consisting of alternating stiff (crystalline) and soft-compliant (disordered) elements. Given the geometrical arrangement of these two alternating phases and the microhardnesses of each of them, the question which arises is how to predict the microhardness of the material. It is known that density, ρ, is a crystallinity parameter which has been successfully correlated to the mechanical behaviour of polymers (yield stress, elastic modulus) (Nielsen, 1954; Reding, 1958; Williamson et al., 1964). Figure 4.6 illustrates the conspicuous increase of microhardness vs density in a double-logarithmic plot for a variety of PE samples. Most revealing are the two straight-line sections which can be associated to the two prevailing mechanisms:

(1) For densities $\rho > 0.92$ g cm^{-3} the deformation modes of the crystals predominate. The hard elements are the lamellae. The mechanical properties are primarily determined by the large anisotropy of molecular forces. The mosaic structure of blocks introduces a specific weakness element which permits chain slip to proceed faster at the block boundaries than inside the blocks. The weakest element of the solid is the surface layer between adjacent lamellae, containing chain folds, free chain ends, tie molecules, etc.

(2) For densities $\rho < 0.92$ g cm^{-3} plastic deformation will probably be dominated by preferential compression of disordered molecular regions. Resistance to deformation occurs largely due to bond rotation. In this case work has to be done against the steric hindrance to rotation.

Extrapolation of the H values for $\rho = 1$ and $\rho = 0.86$ g cm^{-3} yields the limiting values for an ideal PE crystal ($H_c \sim 150$–180 MPa) and an ideal PE amorphous matrix ($H_a \sim 1$ MPa), respectively. It is noteworthy that the extrapolated value obtained for H_c in PE almost coincides with the theoretical value of ultimate shear stress τ_y given in Table 2.1. The experimental H values given in the literature evidently correspond to materials with ρ mostly deviating from unity.

4.2.2   Effect of crystallinity on microhardness

As pointed out above, the semicrystalline polymer can be considered as a two-phase composite of amorphous regions sandwiched between hard crystalline lamellae (Fig. 4.2(a)). Crystal lamellae (ℓ_c) are normally 10–25 nm thick and have transverse dimensions of 0.1–1 μm while the amorphous layer thickness, ℓ_a, is 5–10 nm. As mentioned in the previous section, melt-crystallized polymers generally exhibit a spherulitic morphology in which ribbon-like lamellae are arranged radially in the polycrystalline aggregate (Bassett, 1981). Since the indentation process involves plastic yielding under the stress field of the indenter, microhardness is correlated to the modes of deformation of the semicrystalline polymers (see Chapter 2). These

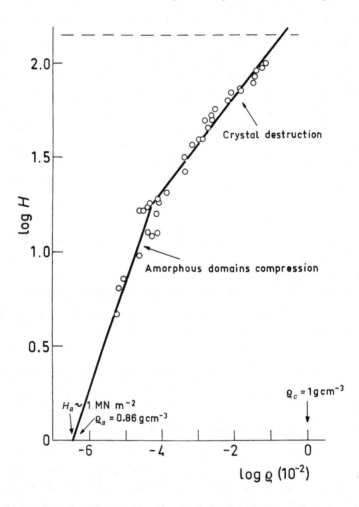

Figure 4.6. Log–log plot illustrating the microhardness H dependence of density ρ for PE samples crystallized from the melt. The plot yields two straight-line sections which can be ascribed to two preferential deformation modes: (top) crystal destruction and (bottom) compression of amorphous domains. (From Baltá Calleja, 1985.)

involve, at small strains, shearing motions of lamellae and lamellar separation (Peterlin, 1987). Beyond the yield point, irreversible deformation processes take place, including lamellar fracture, microfibrillation, and so on.

In dealing with the hardness of semicrystalline polymers it has long been recognized that the following general empirical relationship (parallel model) holds (Baltá Calleja et al., 1981; Baltá Calleja, 1985):

$$H = w_c H_c + (1 - w_c) H_a \qquad (4.3)$$

where H_c and H_a are the intrinsic microhardnesses of the crystalline and the amorphous phases, respectively, and w_c is the volume fraction of crystalline material.[1] In the simple case of flexible polymers, like PE or poly(ethylene oxide) (above T_g), with rubbery amorphous layers (when indentation is done at $T_g < T < T_m$), $H_a \ll H_c$ and we are led to:

$$H \simeq w_c H_c \qquad (4.4)$$

This equation assumes that the semicrystalline polymer is a two-phase system and that the microhardness (yield) is due to plastic deformation taking place only in the crystalline regions. However, for polymers like PET or PEEK, when $T_g > T$ the microhardness of the amorphous phase $H_a \neq 0$ (Deslandes et al., 1991).

A typical good linear relationship between the microhardness and degree of crystallinity has been observed for melt-crystallized PP (Martínez-Salazar et al. 1988). For a series of i-PP samples crystallized within a wide range of crystallinity values the microhardness varies with crystallinity w_c according to eq. (4.3):

$$H = w_c H_{PP}^c + (1 - w_c) H_{PP}^a \qquad (4.5)$$

where H_{PP}^c, and H_{PP}^a are, respectively, the microhardnesses of the crystalline and amorphous phases. In order to derive the limiting values, H_{PP}^c and H_{PP}^a, the plot of H against w_c according to eq. (4.5), can be used (Fig. 4.7). This result characterizes the intrinsic additivity behaviour of both the crystalline and amorphous phases within PP. By extrapolation of H to $w_c = 0$ and $w_c = 1$ one obtains straightforwardly $H_{PP}^a = 30$ MPa and $H_{PP}^c = 116$ MPa, respectively. Knowledge of these quantities allows the immediate determination of H for a given w_c.

A similar linear relationship between H and w_c has also been obtained for other polymers, such as PET (Vanderdonckt et al., 1998) and nylon 6 (Krumova et al., 1998) annealed at various temperatures. The extrapolation of the linear plot of H vs w_c allows one to estimate again the H values for the completely amorphous ($H_a^{PET} \sim 120$ MPa; $H_a^{PA} \sim 50$ MPa) and for the fully crystalline polymers ($H_c^{PET} \sim 400$ MPA; $H_c^{PA} \sim 280$ MPa). In the case of nylon 6 this extrapolation is of particular inetrest because the material is not available in the

---

[1] The most frequently used techniques (X-ray diffraction, density and calorimetry) give the degree of crystallinity as a mass fraction (Gedde, 1995). Conversion of volume fraction crystallinity into the mass fraction crystallinity w_c is straightforwardly given by multiplication of the former with the factor ρ_c/ρ, ρ_c being the crystal density.

fully amorphous state for one to perform microindentation experiments. However, in other semicrystalline systems, such as melt-crystallized linear PE (Baltá Calleja, 1985) and high-pressure-crystallized PET (Baltá Calleja *et al.*, 1994), the hardness data may exhibit a manifest deviation from the additivity rule. This is due to the fact that the upper bound, H_c in eq. (4.3), is not always a constant quantity but rather depends on a crystal's thickness, as we shall see in the next section.

Let us next illustrate the effect of crystallinity on microhardness for a polymer with T_g above room temperature. For this purpose the correlation of H and the microstructure of PET, a polyester of typically low crystallinity having a T_g value well above room temperature, was examined. PET can be easily prepared in the form of a glassy amorphous material by quenching from the melt.

By adequately varying the catalyst content, the temperature, time and the rate of crystallization, materials with a wide range of crystallinities, spherulitic morphologies, different crystal thickness and various levels of crystal distortions have been prepared. The influence of these structural parameters on the microhardness and the detection of specific changes after controlled crystallization are of special interest.

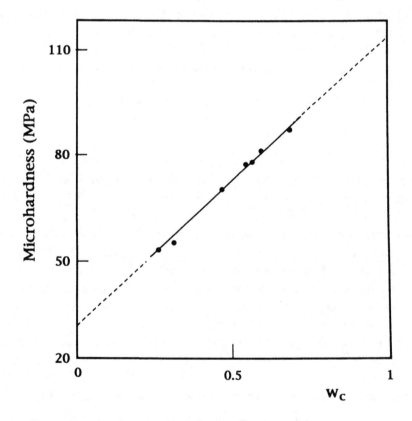

Figure 4.7. Microhardness H of melt-crystallized PP as a function of weight per cent crystallinity w_c. (From Martínez-Salazar *et al.*, 1988.)

All the samples investigated were crystallized from the glassy state for different times at different temperatures. To explain the results obtained, it is necessary to distinguish between two types of morphologies depending on the crystallization conditions: (*a*) structures in which the growth of spherulites is not completed, obtained by interrupting the primary crystallization after different times of crystallization at a constant temperature (Fig. 4.8(*a*)); and (*b*) structures in which such a growth is completed, obtained by crystallization at different temperatures after primary crystallization was completed (Fig. 4.8(*b*)) (Santa Cruz *et al.*, 1991).

In order to obtain samples in which the growth of spherulites is not completed one can crystallize amorphous PET at 117 °C for different times. Spherulites grow from existing nuclei, gradually filling the volume of the material. Thus, it seems plausible that H should be proportional to the crystallinity. Figure 4.9 supports this contention, showing that H is indeed an increasing linear function of the volume fraction of crystallized spherulites $\Phi = w_c/w_{c_0}$ (w_c being the crystallinity after a given time t, and w_{c_0} the degree of crystallinity at which H levels off when crystallization is completed):

$$H = H_a + k\Phi \tag{4.6}$$

where the intercept, H_a, represents the microhardness of the amorphous glassy material which, of course, has no counterpart in eq. (4.3). It is to be noted that eq. (4.6) applies only for samples prepared in the primary crystallization range. The proportionality constant k in eq. (4.6) is equal to the difference in the microhardnesses of the fully crystallized spherulitic material and the amorphous polymer: $H_{sph} - H_a$. Therefore, eq. (4.6) can also be written as:

$$H = H_{sph}\Phi + H_a(1 - \Phi) \tag{4.7}$$

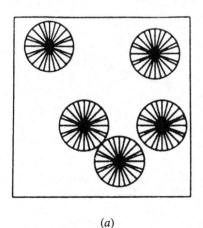

(*a*)   (*b*)

Figure 4.8. Schematic representation of the morphology in spherulitic crystallized PET of: (*a*) incomplete spherulitic growth, (*b*) complete spherulitic crystallization. (From Santa Cruz *et al.*, 1991.)

This equation represents a generalization of the additivity of microhardnesses for high-crystallinity polymers, see eq. (4.3). However, H_{sph} and H_a, instead of describing the crystal and amorphous hardness, now represent the microhardnesses of the spherulites ($H_{sph} \simeq 200$ MPa) and interspherulitic regions ($H_a \simeq 120$ MPa), respectively.

In summary, depending on the structure of the material, microhardness of PET can consequently be characterized as follows:

(a) For the starting amorphous glassy material, $\Phi = 0$ and $H = H_a$.

(b) For samples crystallized from the glassy state where spherulitic growth is incomplete, $0 < \Phi < 1$. In this range, the lamellae thickness $\ell_c = $ constant, as well as the linear degree of crystallinity $w_{c_L} = $ constant, and hence, $H_c = $ constant (Santa Cruz et al., 1991). Therefore, H is directly proportional to the volume occupied by the spherulites. The use of a catalyst induces the formation of a larger number of spherulitic nuclei leading to higher crystallinities, and consequently to a faster H increase.

(c) For samples crystallized from the glassy state in which spherulitic growth is completed, $\Phi = 1$. Here the crystal thickness increases slightly with crystallization temperature T_c, leading to a concurrent small increase of the crystal microhardness H_c. However, since w_{c_L} remains constant with T_c, and the increase in H_c is small, the resulting end effect is that one observes hardly any H variation with T_c. Finally, it is noteworthy that the amorphous domains outside the lamellar stacks do not influence the value of H (Santa Cruz et al., 1991).

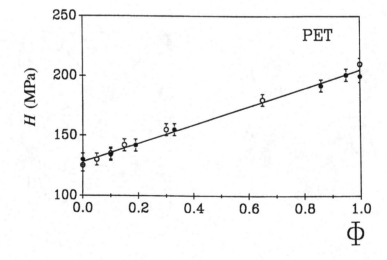

Figure 4.9. Plot of microhardness H of PET as a function of volume fraction of spherulites Φ (eq. (4.6)). For $\Phi = 1$ the spherulites fill the volume of the sample and $H = H_{sph}$. (●) Samples with a Mn catalyst; (○) samples without catalyst. (From Santa Cruz et al., 1991.)

4.2.3 Models to predict hardness of lamellar crystals: role of lamellar thickness

From eq. (4.3) it is clear that w_c controls the microhardness of a polymer. However, the structure of the semicrystalline polymers, as pointed out above, is characterized by stacks of crystalline lamellae. The direct influence of lamellae thickness, ℓ_c, upon microhardness in the case of chain-folded and chain-extended PE samples was soon recognized (Baltá Calleja, 1976, 1985). Popli & Manderkern (1987) and Crist et al. (1989) later provided independent evidence for the strong dependence of yield stress, σ_y, upon ℓ_c. The above authors found that spherulite size does not have a direct influence upon yield in PE. On the basis of a heterogeneous deformation model involving the heat dissipated by the plastically deformed crystals, Baltá Calleja & Kilian (1985) developed an approach to calculate the dependence of microhardness on the average crystal thickness ℓ_c. Figure 4.5 shows lamellar deformation under the indenter which involves the generation of a number of shear planes. In Fig. 4.5 $\langle D_c \rangle$ and ℓ_p represent the average size of the crystallites normal to the chain axis, before and after plastic deformation, respectively (Baltá Calleja & Kilian, 1985). The microhardness of the crystals can then be described by:

$$H_c = \frac{H_c^0}{1 + b/l_c} \tag{4.8}$$

where H_c^0 is the microhardness of an infinitely thick crystal (maximum possible value of energy dissipated through plastic deformation) and b is a parameter related to the basal surface free energy, σ_e, of the crystal and to the energy required for plastic deformation of the crystals, Δh, through formation of a great number of shearing planes. The b parameter is given by:

$$b = 2\sigma_e/\Delta h \tag{4.9}$$

Figure 4.10 shows a plot of H_c as a function of ℓ_c^{-1} for different PE and paraffin samples crystallized from the melt. The calculated continuous curve was derived using $b = 20$ nm. The deviation of the experimental data from the calculated curve is due to the wide range of b values obtained ($b = 6$–26 nm), corresponding to the widely different structures examined (CEPE crystals, chain-folded lamellae, low-density (LD) PE and linear paraffins). Comparison of experimental and calculated H_c data for PE emphasizes the parallel increase in the b parameter derived from mechanical data (14–26 nm) and in σ_e determined thermodynamically (79–91 mJ m^{-2}) for a series of samples with different molecular weights (56 000–307 000) (Baltá Calleja et al., 1990b) (see also Section 4.3).

A new thermodynamic derivation of eq. (4.8) has been proposed making use of a modified Clausius–Clapeyron equation. The derivation of this equation is based on the assumption that plastic deformation involves a partial melting of the polymer crystals (Hirami et al., 1999).

Young (1988) and Crist *et al.* (1989) applied a different approach, describing the dependence of the 'crystal' yield stress, σ_y^0, on ℓ_c in terms of the screw dislocation model for yield. Figure 4.11(a) shows a schematic representation of a laminar PE crystal of thickness ℓ_c under an applied shear stress τ containing a screw dislocation of Burgers vector, **b**, lying a distance ℓ from the edge. The dependence of σ_y^0 upon ℓ_c is given by:

$$\sigma_y^0 \simeq \frac{H_c}{3} = \frac{K}{2\pi e} \exp\left(-\frac{2\pi \Delta G^*}{K \tilde{\mathbf{b}}^2 \ell_c}\right) \tag{4.10}$$

where K is a function of the shear modulus of the crystals and $\tilde{\mathbf{b}}$ and ΔG^* are the Burgers vector and the activation energy of the screw dislocation, respectively. Values of $\Delta G^* \simeq (40\text{–}60)kT$ have been used. Comparison of the Baltá Calleja & Kilian (1985) approach with the Young (1988) dislocation model for yield gives good agreement for oriented PE films (Santa Cruz *et al.*, 1993). However, when using the yield dislocation model, appreciable deviations between calculated and experimental values for isotropic melt-crystallized samples are observed (see

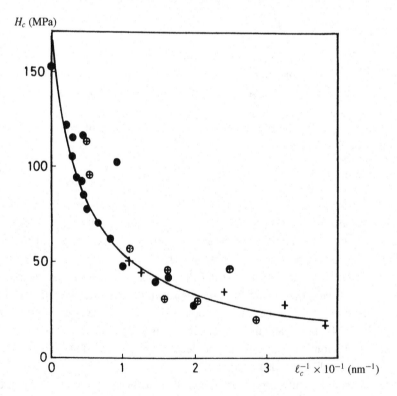

Figure 4.10. Dependence of crystal microhardness, H_c upon lamellae thickness, ℓ_c for a series of PE samples: ($\oplus$) quenched PE; ($\bigcirc$) PE slowly cooled from the melt; and ($+$) melt-crystallized paraffin. (After Baltá Calleja & Kilian, 1985.)

Fig. 4.11(*b*)). It is believed that this is because Young's model was developed for the plastic deformation of crystals in a direction parallel to the molecules (Santa Cruz *et al.*, 1993).

As we have seen above, in order to study the effect of crystal size on the microhardness one has to exclude the influence of the degree of crystallinity (see Section 4.2), i.e. one has to take into account samples with the same crystallinity

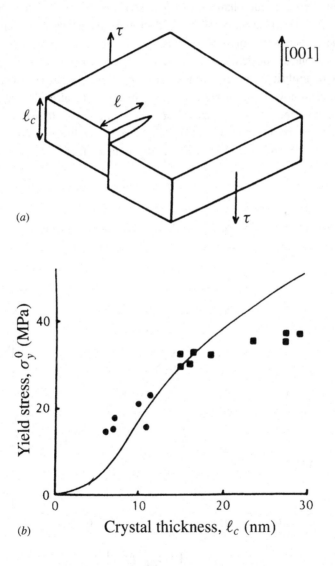

(*a*)

(*b*)   Crystal thickness, ℓ_c (nm)

Figure 4.11. (*a*) Schematic representation of a screw dislocation in a lamellar single crystal of PE. The chain direction is [001]. (*b*) Dependence of yield stress σ_y^0 on crystal thickness ℓ_c for crystals of branched PE (●) and linear PE (■). The continuous line is calculated from eq. (4.10). (After Young, 1988.)

but of different crystal size. Krumova *et al.* (1998) investigated polyamide 6 (PA6) drawn and annealed bristles with w_c(DSC) values between 30 and 35% and crystal sizes $D_{(200)}$ of between 33 and 106 Å. The effect of crystal size on microhardness is shown in Fig. 4.12. Extrapolation of the straight line yields for $1/D_{200} = 0$ the microhardness of PA6 (composed of roughly 70% amorphous phase and 30% crystalline phase) for infinitely large crystals. The microhardness obtained was 153 MPa. This means that a completely crystalline PA6 comprising infinitely large crystals should have a microhardness of about $H_c^0 = 460$ MPa. It is important to note that the extrapolation of H vs w_c also gives the H values of completely crystalline species but for polymer comprising crystallites with finite sizes and not infinitely large ones. For this reason the H_c^0 value of 460 MPa can be considered as an ideal (equilibrium) microhardness of PA6, similarly to the ideal melting temperature (Krumova *et al.*, 1998). The plot in Fig. 4.12 also illustrates that to demonstrate the effect of crystal size on H it is not necessary to use the lamellae thickness ℓ_c as determined by SAXS; the crystal size in any direction D_{hkl} evaluated by means of WAXS which changes with the crystallization conditions may also be used.

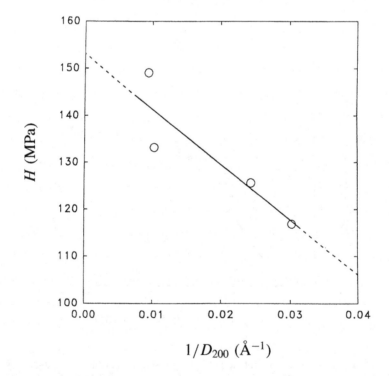

Figure 4.12. Plot of microhardness H vs reciprocal crystal size $1/D_{200}$, for PA6 samples with the same crystallinity. (After Krumova *et al.*, 1998.)

The strong effect of the crystal size on the crystal microhardness allows one to evaluate the 'ideal' H_c^0 value by means of extrapolation of ℓ_c or D_{hkl} to infinitely large crystals.

We may rewrite eq. (4.8) as

$$\frac{1}{H_c} = \frac{1}{H_c^0}\left(1 + b\frac{1}{\ell_c}\right) \tag{4.11}$$

According to eq. (4.11), H_c^{-1} increases linearly with ℓ_c^{-1} at constant b. The ordinate intercept of a H_c^{-1} vs ℓ_c^{-1} plot gives $1/H_c^0$, and the slope is proportional to b. Let us examine next the plot of eq. (4.11) in case of chain-folded PE and CEPE samples (Fig. 4.13) (Flores et al., 1999). The value $H_c^0 = 170$ MPa for $\ell_c^{-1} = 0$ has been also included in this plot. Figure 4.13 shows that the data for chain-folded linear PE samples can be fitted to a straight line yielding $b = 21$ nm (the intercept agrees with $1/170$ MPa^{-1}) (Baltá Calleja & Kilian, 1985). The value of b obtained is consistent with other values of b reported for slowly crystallized PE samples of similar molecular mass (Bayer et al., 1997; Baltá Calleja et al., 1995). Data for CEPE are also presented in Fig. 4.13. Here the ℓ_c values correspond to the weighted average of the ℓ_c values derived from gel permeation chromotography (GPC) measurements. A fit of the data of the high-pressure crystallized (or annealed) samples (assuming an ordinate intercept of $1/170$ MPa^{-1}) yields $b \approx 85$ nm.

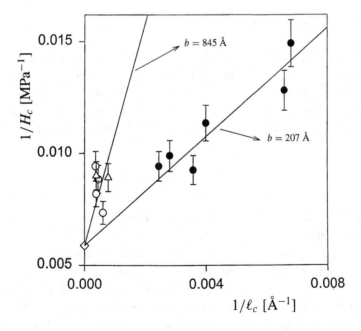

Figure 4.13. Plot of H_c^{-1} vs ℓ_c^{-1} for chain-folded (●) and chain-extended PE (○, △). (After Flores et al., 1999.)

The data of Fig. 4.13 show that the b values for CEPE crystals are notably larger than for chain-folded lamellae. What can be the reason for this difference? Let us recall first that, according to eq. (4.11), for a given average lamellar thickness, ℓ_c, the b determines the microhardness of the lamellae crystallites, H_c, and second that b is, in addition, directly proportional to the basal surface free energy. Previous investigations on CEPE and chain-folded crystals have shown a common reduced plot of the fractional change in crystal thickness against undercooling. Consequently, the basal surface free energies of the chain-extended and the chain-folded crystals are similar. This suggests that the higher b value observed for chain-extended lamellae with respect to chain-folded ones is presumably due to a lower Δh value (see Fig. 4.13). The interlamellar regions in oriented CEPE have been shown to exhibit a higher mass deficit per unit area than in chain-folded polymer (Bassett & Carder, 1973). This fact, together with a reduction of the entanglements' density in the amorphous regions within CEPE lamellae, could explain the lower energy required for plastic deformation of the CEPE lamellae with respect to the chain-folded material (Flores *et al.*, 1999).

4.2.4   Relationship between crystal hardness and melting temperature

The analogy between eq. (4.8) and the Thomson–Gibbs equation:

$$T_m = T_m^0 \left(1 - \frac{b^*}{\ell_c} \right) \tag{4.12}$$

where T_m^0 is the equilibrium melting point and $b^* = 2\sigma_e/\Delta h_f$ (Δh_f being the melting enthalpy), has been highlighted before (Baltá Calleja *et al.*, 1990). While the crystal hardness depression due to the finite thickness of the lamellae is given by the ratio b, the melting point depression is characterized by an equivalent ratio b^* which is related to the equilibrium properties of the crystals. By combining eqs. (4.11) and (4.12) we obtain:

$$\frac{1}{H_c} = -\frac{b}{b^* H_c^0 T_m^0} T_m + \frac{1}{H_c^0} \left(1 + \frac{b}{b^*} \right) \tag{4.13}$$

Equation (4.13) shows that the reciprocal value of the crystal hardness is related to the melting temperature of the crystals.

According to eq. (4.13), the H_c^{-1} vs T_m plot for different polymeric materials will fit a straight line provided the b/b^* ratio is constant. Figure 4.14 shows the plot of H_c^{-1} vs T_m for chain-folded PE and CEPE samples, for high-pressure crystallized PET and for atmospheric-pressure crystallized poly(ethylene oxide) (PEO) samples. From the linear regression fit for the H_c^{-1} vs T_m data of PEO, PE and PET (Fig. 4.14), values of $H_c^0 = 170$ MPa for PE, 150 MPa for PEO and 392 MPa for PET are obtained (assuming $T_m^0 = 558$ K for PET, $T_m^0 = 414$ K for PE and $T_m^0 = $

342 K for PEO) in agreement with previous results (Baltá Calleja & Santa Cruz, 1996; Baltá Calleja et al., 1994). The b/b^* ratio derived from the slope of the plots is around 2 for PET, 44 for PE and 95 for PEO. This result suggests that for flexible polymers the energy required for plastic deformation of the crystals is much lower than the melting enthalpy. As the chain stiffness increases, the b/b^* ratio seems to decrease as a consequence of a higher energy required for crystal deformation.

4.3 Microhardness and surface free energy

Polymer surfaces is a field of increasing interest to both basic and applied research (Eisenriegler, 1993). The aim of this section is to show that microhardness is directly related to surface free energy and, therefore, to the degree of polymer perfection at polymer surfaces and interfaces. Studies have revealed that the morphology (crystal thickness and size of the interlamellar regions) of the polymer nanostructure are the main factors determining the microhardness (Baltá Calleja et al., 1997). (See also Section 4.2.3.) The hardness-derived parameter

$$ b = \ell_c \left(\frac{H_c^0}{H_c} - 1 \right) \tag{4.14} $$

offers a measure of the hardness depression due to the finite thickness of the crystal lamellae (see also eq. (4.8)). Furthermore, b is directly related to the surface free energy σ_e through eq. (4.9).

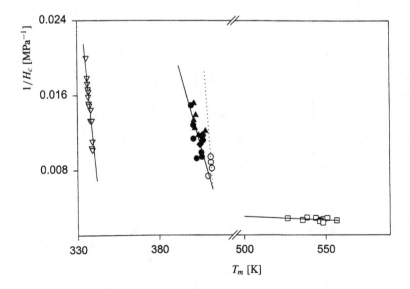

Figure 4.14. Plot of $1/H_c$ vs T_m for chain-folded PE ($\bullet$), PEO ($\triangledown$), CEPE ($\circ$) and PET ($\square$). (From Flores et al., 1999.)

It will be shown that changes in b in various polymers allow the estimation of changes in σ_e which are related to the level of defects incorporated at the crystalline–amorphous interphase.

4.3.1 The role of entanglements

In the case of melt-crystallized PE samples (Baltá Calleja *et al.*, 1990a) the increase in b derived from H_c (eq. (4.14)) with molecular weight has been shown to parallel the increase in surface free energy σ_e derived from DSC experiments. Figure 4.15 illustrates the good fit between experimental and calculated (from eq. (4.14)) data for b as a function of molecular weight. The increase in σ_e has been interpreted in terms of an increase in the number of defects and molecular entanglements located on the surface boundary of the lamellar crystals. As we shall see in Chapter 5, variations of H with composition in PE/PP gel blends have also been discussed in the light of changes occurring in the defective boundary of the lamellar crystals (Baltá Calleja *et al.*, 1990a). In previous studies of melt-crystallized PE the molecular weight range investigated was sufficiently high (56 000–307 000) to allow one to assume that the chain ends had no effect on σ_e. The variations in σ_e were mainly ascribed to the increasing number of defects and molecular entanglements located at the crystal surface. Later investigations support this view and emphasize the correlation between crystal microhardness and the average distance between stable entanglements (Bayer *et al.*, 1997).

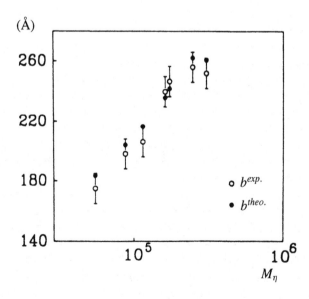

Figure 4.15. Experimental (from eq. (4.14)) and calculated data for b as a function of molecular weight M_η for melt-crystallized linear PE. (After Baltá Calleja *et al.*, 1990.)

On the other hand, microhardness studies of short paraffin crystals reveal the influence of the molecular packing and hence of the chain ends on b (Ania *et al.*, 1986).

4.3.2 Influence of chain ends

It is important, as regards the surface free energy, to investigate the microhardness of materials in the molecular weight range corresponding to the transition from straight-chain crystallization of oligomers to the folded-chain crystallization of polymers. The crystallization of melt-crystallized PEO and the characterization of its lamellar structure by SAXS were the objective of an early study in which the effect of molecular weight on fold length was thoroughly investigated (Arlie *et al.*, 1966, 1967). In the present work we wish to illustrate the influence of chain ends and chain folds, on both the surface free energy as derived from DSC data using eq. (4.12) and the b parameter using eq. (4.14).

The crystal thickness ℓ_c and the surface free energy of the PEO crystals may be changed simultaneously by changing the molecular weight of the samples.

Table 4.1 lists the density ρ, degree of crystallinity from density measurements $w_c(\rho)$, degree of crystallinity from DSC (w_c(DSC)), long period L, lamellae thickness ℓ_c, microhardness H, crystal microhardness H_c, melting temperature T_m, surface free energy σ_e and number of chain folds per molecule n for the PEO samples investigated as a function of molecular weight, M_n. As expected, one sees a rapid increase of w_c, L and ℓ_c for low molecular weights, and a levelling off tendency at $M_n \sim 2 \times 10^3$. Note the good agreement obtained between $w_c(\rho)$

Table 4.1. *Experimental values for the density, ρ, crystallinity measured by density, $w_c(\rho)$, crystallinity measured by DSC, w_c(DSC), long period, L, crystal thickness, ℓ_c, microhardness, H, crystal hardness, H_c, melting temperature, T_m, surface free energy, σ_e, and the number of chain folds per molecule, n, for PEO samples with different molecular weights. (From Baltá Calleja & Santa Cruz, 1996.)*

M_n	ρ (g cm^{-3})	$w_c(\rho)$	w_c (DSC)	L (Å)	ℓ_c (Å)	H (MPa)	H_c (MPa)	T_m (K)	σ_e (erg cm^{-2})	n
1 000	1.2064	0.804	0.78	73	59	3	3.2	310.3	65.7	0
1 300	1.2132	0.870	0.85	80	70	4	4.5	316.1	63.6	0
2 000	1.2159	0.896	0.87	132	118	9	10.0	326.9	62.4	0
3 500	1.2161	0.898	0.88	140	126	23	25.4	331.3	47.4	1
6 700	1.2169	0.906	0.91	155	140	53	58.0	335.0	34.3	2
13 000	1.2184	0.921	0.91	165	152	59	63.7	663.0	31.7	4
17 000	1.2183	0.895	0.89	147	135	43	47.7	333.4	40.7	6

and w_c(DSC). For $M_n > 2 \times 10^4$ a slight crystallinity decrease is observed. This decrease may be due to the increase in viscosity with increasing molecular weight which hampers crystallization from the melt. Further, for low molecular weights the chains are totally extended in the crystals, $n = 0$ (Table 4.1), but with increasing molecular weight the number of chain folds in the crystals increases (Table 4.1).

Figure 4.16(a) shows the variation of H and H_c for PEO as a function of molecular weight. For low molecular weights the H values are extremely low ($H \approx 3$ MPa). However, H increases with M_n up to a maximum of $H \sim 65$ MPa for $M_n \sim 2 \times 10^4$ after which the hardness decreases again. The variation of H

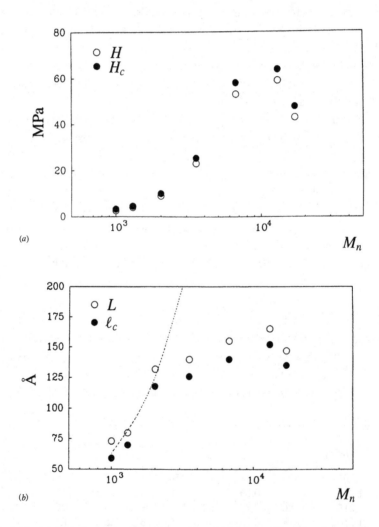

Figure 4.16. (a) Variation of hardness, H, and crystal hardness, H_c, of melt-crystallized PEO as a function of molecular weight. (b) The long period L, lamellar thickness ℓ_c and extended-chain length (dashed line) of the above PEO samples as a function of molecular weight. (After Baltá Calleja & Santa Cruz, 1996.)

with M_n approximately follows the changes observed in crystallinity and crystal thickness (see Table 4.1 and Fig. 4.16(b)).

It should be noted that with increasing molecular weight there is a transition from extended-chain to folded-chain crystals. This transition must dramatically affect the surface free energy of the crystals. As pointed out by Buckley & Kovacs (1976) and Baltá Calleja et al. (1969), when the crystals have extended chains there is a high proportion of hydrogen bond formation between the –OH groups at adjacent chain ends.

Let us now discuss the molecular weight dependence of b. The results in Table 4.1 show the substantial variation of σ_e with M_n, corresponding to the transition from straight-chain structures ($n = 0$) to the chain-folded morphology ($n \neq 0$) in which chain ends are progressively replaced by chain folds in the crystal surfaces ($n \neq 0$). It is interesting to note that the values of σ_e obtained for the highest-molecular-weight sample are in accordance with the theoretical data derived by Buckley and Kovacs (1976).

The decrease of surface free energy with increasing chain folds can be interpreted by assuming that hydrogen bonds at the surface of the crystals are progressively replaced by chain folds. However, for the highest-molecular-weight sample with $n = 6$, where no hydrogen bond formation is expected, the value of σ_e shows an increase with respect to the samples with $n = 2$ and $n = 4$. One could speculate that this result is in accordance with a rise in the number of defects and molecular entanglements with increasing molecular weight, as observed previously in PE samples.

Table 4.1 shows, in addition, that in crystalline lamellae with extended chains ($n = 0$), the surface free energy has a nearly constant value which corresponds to a straight line. With increasing molecular weight the number of chain folds increases and the surface free energy diminishes, leading to two families of curves. One now sees that the similar variations of T_m and H_c with $1/\ell_c$ are due to the same variation in σ_e of the crystals with increasing molecular weight.

Let us next compare the variation of the b and b^* parameters with molecular weight and analyse the relationship, b/b^*, between them. The initial decrease in b^* is confirmed by a similar decrease of b with increasing molecular weight.

It is noteworthy that the ratio b/b^* (Table 4.2) does not remain constant as in the case of the samples crystallized at different temperatures; it diminishes with increasing molecular weight. The reason for this is that b diminishes faster with M_n than does b^*. Since σ_e is the same for both b and b^*, the difference in the variation of these parameters with M_n must originate in the different variations of Δh and Δh_f with M_n. As Δh_f should have a constant value which is independent of M_n the observed variation in b/b^* must be due to changes in Δh. One possible explanation for this effect may lie in the occurrence of tie molecules between adjacent crystals. Tie molecules may be expected to appear with increasing molecular weight, thus providing a reinforcement of the material which could contribute to an increase in the energy of destruction (Δh), and lead consequently to a b decrease. On the other hand, the tie molecules should not affect the thermodynamical parameter b^*. With these considerations in mind one can conclude that there are two different decreasing contributions to b in the molecular weight range between 10^3 and 10^4: first a contribution due to an increasing Δh owing to the presence of tie molecules and, second, a contribution due to the increase in the number of chain folds with increasing M_n.

4.4   Influence of molecular weight on microhardness through physical structure

In Chapter 1 it was mentioned that there is no obvious reason for the effect of molecular weight on microhardness, i.e. it is not easy to predict the type of this dependence. Probably, this is because the molecular weight affects the various factors which actually determine H (such as crystallinity, crystal size, crystal perfection, etc.) in different ways, sometimes even in opposing directions. In the foregoing section the influence on microhardness of the degree of perfection at the polymer interface was discussed. In this section, the effect of lamellar thickness and the degree of crystallinity on microhardness will be illustrated on different polymers following the dependence of the H-determining factors on molecular weight.

The microhardness of LDPE fractions and of high-density polyethylene (HDPE) samples has been studied as a function of molecular weight in the range of about 2×10^4 up to 4×10^6 (Baltá Calleja *et al.*, 1997). Details of the lamellar structure were determined by TEM. The observed decrease of microhardness with increasing molecular weight is mainly due to the increase in thickness of the interlamellar layers (i.e. a decrease of crystallinity). After chemical treatment with chlorosulphonic acid and with OsO_4, samples show a drastic increase in microhardness.

Table 4.2. *Calculated values for b and b* and the ratio between them as a function of molecular weight, M_n for melt-crystallized PEO.*

M_n	b (Å)	b^* (Å)	b/b^*
1 000	2796	5.4	501
1 300	2263	5.2	435
2 000	1652	5.1	323
3 500	618	3.9	158
6 700	222	2.8	79
13 000	206	2.6	79
17 000	290	3.4	85

The kinetics of the OsO_4 treatment as revealed by microhardness measurements has been examined. The increase in microhardness has been explained in terms of the large reduction in molecular mobility of the amorphous, interlamellar layers, Chlorosulphonation produces an initial microhardening of the amorphous phase, in which reaction time, temperature and the molecular weight play an important role. The OsO_4 reaction induces an additional microhardening at the surface of crystalline lamellae. Results reveal that H_a increases, after the above chemical reactions, from nearly zero up to values of 300 MPa (Baltá Calleja et al., 1997).

The influence of molecular weight on the mechanical properties of i-PP and i-PP blended with ethylene-propylene copolymers has also been investigated by means of the microhardness technique (Flores et al., 1998). The microhardness of i-PP was shown to decrease slightly with molecular weight, in the range of molecular weights investigated (between 200 000–900 000). The H decrease correlates to a loss of crystallinity as the average molecular weight increases. On annealing, the mechanical properties are enhanced as a consequence of an increase in both the degree of crystallinity and the crystalline lamellar thickness. The inclusion of ethylene–co-propylene rubber (EPR) particles in the i-PP matrix softens the material owing to the rubber-like nature of the EPR component. In the study by Flores et al. (1998) a value of $H_c^0 = 230$ MPa for the α-form of i-PP was proposed.

It seemed necessary to complement the studies described above on the effect of molecular weight on microhardness by studying a range of intermediate molecular weights. For this purpose samples of PET with different molecular weights (in the range 10 000–120 000) were prepared by means of solid-state post-condensation (Fakirov, 1990). The primary aim of the study was to distinguish between the contributions of the crystallinity and molecular weight to the measured hardness, as both quantities develop simultaneously during the thermal treatment (Vanderdonckt et al., 1998).

PET was annealed in vacuum at different temperatures (190–260 °C) for different times (10 min–24 h) in order to obtain samples with a wide range of molecular weights (10 000–120 000). Short annealing times result in a twofold decrease in molecular weight due to hydrolytic decomposition. However, long annealing times give rise to a substantial molecular weight increase. It was found that microhardness rises linearly with the degree of crystallinity obtained during upgrading of molecular weight and its extrapolation leads to H values of completely crystalline PET ($H_c^{PET} = 405$ MPa) for samples of conventional molecular weight and of 426 MPa for samples with a molecular weight higher than 30 000. An increase in molecular weight for two sets of samples with different ranges of degree of crystallinity (39–42% and 45–53%) also caused a slight increase of H as can be concluded from Fig. 4.17.

How can the relationship between molecular weight and microhardness seen on Fig. 4.17 be explained?

Obviously the effect of molecular weight on microhardness is transferred via the physical structure (crystallinity and crystal thickness) arising during treatment. By starting from the very general knowledge about the effect of molecular weight on structure in polymers one would not expect species with the highest molecular weights to be distinguished by the most perfect structures, which result in the highest values of H. This apparent contradiction is related to a peculiarity in the preparation of PET samples with higher molecular weight. Since this is done by means of long annealing times at very high temperatures which favours the formation of a rather perfect structure, the samples with the highest molecular weights are also expected to have the most perfect structure. This suggestion is supported by the data displayed in Fig. 4.17. The PET samples distinguished by higher H have crystallinities around 50%, while the set with lower H have crystallinities of only around 40%. The well expressed dependence of H on molecular weight in the second case is related with changes in the lamellar thickness. All the samples in this series have w_c in the range 39–42%. However, with increasing T_a and t_a, when PET samples with the highest molecular weights are prepared, a systematic increase in ℓ_c (from 35 to 56 Å) is observed (Vanderdonckt *et al.*, 1998).

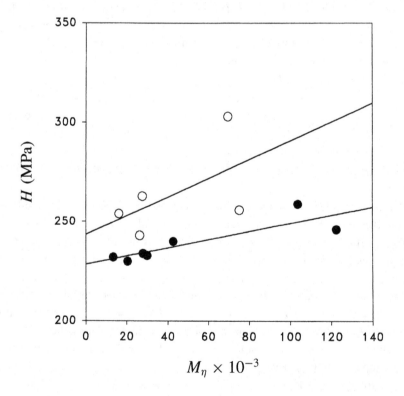

Figure 4.17. Dependence of the microhardness H on the molecular weight M_η of two sets of PET films with differing degrees of crystallinity w_c: (●) $w_c(\rho)$ between 39 and 42% (○) $w_c(\rho)$ between 43 and 53%. (After Vanderdonckt *et al.*, 1998.)

4.5 Transitions in crystalline polymers as revealed by microhardness

4.5.1 Polymorphic transitions

The increasing variation of the unit cell dimensions of polymers (PE, PEEK) with temperature leads to a decrease in microhardness (Deslandes *et al.*, 1991; Baltá Calleja *et al.*, 1985). For example, in linear PE, H decreases from 90 to 70 MPa over the temperature range 20–70 °C. The better packing of the chains and, therefore, the changes in cohesive energy seem to play a role in determining the microhardness. Since microhardness is related to the packing of the chains in the crystals, polymorphic changes may be expected to influence the strength of the intermolecular forces and, hence, the microhardness.

The influence of polymorphism on hardness can be illustrated using the case of PP. *i*-PP can crystallize into several polymorphic forms. Microhardness measurements have been carried out on samples prepared by the temperature-gradient method of Fujiwara & Asano (Fujiwara, 1968; Asano & Fujiwara, 1978), which allows a distinct separation of the α (monoclinic) and β (hexagonal) phases within the sample. Results show that H decreases from 111 MPa to 90 MPa when passing from the α to the β phase (Baltá Calleja *et al.*, 1988). The H decrease is partially connected with a crystallinity decrease from 0.72 to 0.62. However, according to eq. (4.7) the microhardness of each phase depends not only on crystallinity but also on the crystal microhardness, H_c, itself. Thus, we can write for both crystalline phases:

$$H^\alpha = w_c^\alpha \, H_c^\alpha + (1 - w_c^\alpha) \, H_a \qquad (4.15)$$

$$H^\beta = w_c^\beta \, H_c^\beta + (1 - w_c^\beta) \, H_a \qquad (4.16)$$

where w_c^α and w_c^β represent the degree of crystallinity of the α and β phases, respectively.

If we take a value of $H_a \simeq 30$ MPa (Martínez-Salazar *et al.*, 1988) for the amorphous phase of *i*-PP and solve eqs. (4.15) and (4.16), values of $H_c^\alpha = 143$ MPa and $H_c^\beta = 119$ MPa are obtained. The former value of 143 MPa fits well with the *ab initio* calculation for the α phase (Baltá Calleja *et al.*, 1988). In conclusion, the determination of microhardness is shown to be a technique capable of detecting polymorphic changes in polymers. Further examples of polymorphic crystal–crystal transitions induced by external field (stress or strain) are given in Chapter 6.

4.5.2 Curie transition

Evidence for ferroelectric-to-paraelectric phase transformations in random copolymers of VF_2 and trifluorethylene (F_3E) as revealed by WAXS and SAXS has been reported (Tashiro *et al.*, 1981, 1984; Lovinger *et al.*, 1982, 1983). The conformation of these copolymers at low temperatures in their ferroelectric phase is the same as in β-PVF$_2$ (Lovinger, 1985) (see Fig. 4.3), i.e. chains in a planar all-*trans* conformation packed in a pseudo-hexagonal lattice so that the dipoles are parallel to the b axis. Above the Curie temperature this conformation changes into a disordered sequence of conformational isomers $(TG, T\tilde{G}, TT)$ (Lovinger *et al.*, 1983) as a result of the introduction of gauche bonds, which rotate around the chain axis so that the unit cell is no longer polar. On cooling to room temperature the high-temperature phase does not fully return to the original ferroelectric low-temperature phase but is partly changed into a low-temperature non-polar phase. Based on X-ray, IR and Raman results Tashiro & Kobayashi (1985) visualized this phase as a kind of superlattice structure consisting of an aggregation of polar microdomains, with all-*trans* zigzag chains, which are in a considerably disordered state contributing to the observed broadening and diffuseness of the X-ray scattering patterns. For this reason it seemed interesting to examine the behaviour of the mechanical properties of copolymers of VF$_2$ and F$_3$E through the Curie transition region as a function of temperature by measuring the microhardness (Baltá Calleja *et al.*, 1992).

Figure 4.18 shows the H variation obtained as a function of increasing (Fig. 4.18(a)) and decreasing (Fig. 4.18(b)) temperature for three copolymers with molar concentrations of 60/40, 70/30 and 80/20 VF$_2$–F$_3$E. In all cases log H was observed to decrease linearly with increasing temperature. The onset of the Curie transition can be clearly identified by a bend in the log H *vs* T plot. On cooling (Fig. 4.18(b)) H increases again with decreasing temperature, and the point at which the Curie transition ends can again be identified by the change in slope of the plot. It is noteworthy that the onset of the Curie transition temperature, as determined from the measurement of microhardness, is lower during cooling than during heating. This is due to the thermal hysteresis of the exothermic T_{Curie} peaks which appear after solidification at lower temperatures (Koga & Ohigashi, 1986).

Figure 4.19 illustrates the relationship between the Curie transition temperature as determined from microhardness and as measured by calorimetry. Only for the 80/20 composition is there an uncertainty in the correlation, due to the overlapping of the Curie temperature with the melting point.

Most interesting is the fact that H measured at room temperature (Fig. 4.18) appreciably increases with VF$_2$ content. Since the fraction of ferroelectric crystals is nearly constant ($w_c \simeq 0.5$–0.6) for the various compositions the increase in microhardness at low temperature can be explained on the basis of the structural characteristics of the ferroelectric phases of β-PVF$_2$ and its copolymers. In each case the molecular conformation is the same. The intermolecular lattice structure, chain packing and dipolar alignment are also the same. The major difference between PVF$_2$ and its copolymers lies in the spacing of their intermolecular lattices. In addition, in β-PVF$_2$ the dipole moments of the individual segments with a *trans* conformation reinforce each other, while in the case of its copolymers the dipole moments from the $-CF_2-$ groups are partially compensated by the moments of the $-CFH-$ groups which are randomly distributed along the opposite side of the chain.

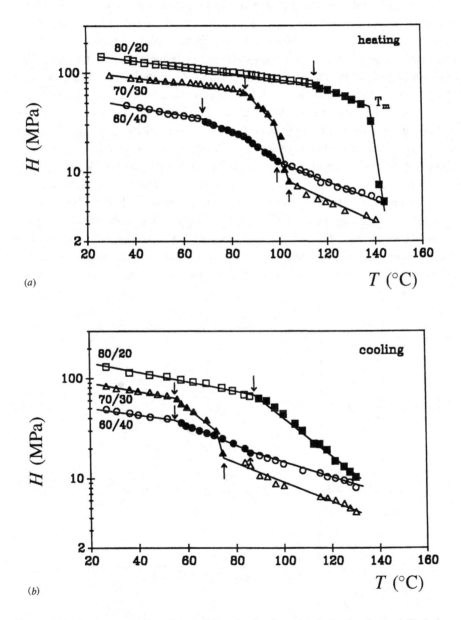

Figure 4.18. Temperature dependence of the microhardness (a) during heating and (b) during cooling of the 60/40 (○), 70/30 (△), 80/20 (□) VF_2–F_3E copolymers. The filled symbols denote the Curie transition region between ferroelectric (low T) and paraelectric (high T) and paraelectric (high T) behaviour. (From Baltá Calleja, Santa Cruz et al., 1992.)

One can conclude that the microhardness measurement is an appropriate technique for accurately detecting ferroelectric-to-paraelectric phase changes in VF_2–F_3E copolymers. At room temperature the increase in microhardness as a function of increasing VF_2 polar sequences has been correlated with the contraction of the unit cell and the parallel reduction in the number of dipoles, due to a decrease of bulky F_3E units within the ferroelectric crystals. The microhardness of these materials when measured over a range of temperatures shows the following behaviour. At low temperatures (below T_{Curie}) the microhardness of ferroelectric crystals decreases exponentially as a function of temperature, and is characterized by a thermal softening coefficient $\beta_F \simeq 0.01$ K^{-1} similar to that found in other crystalline polymers in the same temperature range. At the Curie temperature the microhardness shows a sharp decrease on a plot of $\log H$ vs T, which characterizes a transition from the ferroelectric to the paraelectric phase involving a complex molecular reorganization within the crystals. At higher temperatures (above T_{Curie}) the paraelectric crystals present a much faster exponential H decrease with T which is characterized by a coefficient $\beta_P \simeq 0.2$–0.45 K^{-1} similar to that of amorphous polymers above T_g (see Chapter 3), suggesting the existence of a 'liquid-crystalline' state. Finally, a study of the microhardness of these materials during heating and cooling cycles revealed the existence of a hysteresis behaviour of the microhardness with temperature which is caused by a shift in T_{Curie} towards lower temperatures during the cooling process.

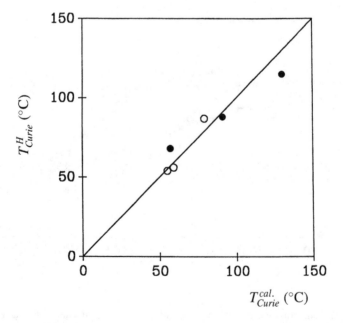

Figure 4.19. Correlation between the Curie temperatures T_{Curie} (●) determined from microhardness and (○) measured by calorimetry. (After Baltá Calleja et al., 1992.)

4.6   Crystallization kinetics as revealed by microhardness

The crystallization of amorphous polymers (e.g. PET from the glassy state) has been examined by measuring the microhardness in 'real time' at different temperatures (Baltá Calleja *et al.*, 1993). We highlighted the occurrence of some transitions near T_g that can be evidenced by means of microhardness measurements *in situ*. Figure 4.20 illustrates the kinetics of the microhardness changes during crystallization of physically aged PET as a function of time for various temperatures. For samples without a nucleating agent, one single rate of increase at each crystallization temperature, T_c, was observed until a final plateau was reached (Fig. 4.20(*a*)). Most interesting is the fact that the initial variation of hardness with time (Fig. 4.20(*b*)) shows a maximum after 8–9 min when measured at both 105 °C and 110 °C, just before the onset of primary crystallization. Fresh samples measured immediately after quenching do not show such H maxima. One explanation of these transitions revolves around a possible restriction of segmental motions involving a time-dependent molecular ordering, giving rise, in turn, to a hardening effect. The hardening of the material vanishes after 5–10 min for PET without a nucleant. For PET with a nucleating agent the hardening occurs after longer times (Baltá Calleja *et al.*, 1993). This transition could be visualized as a precrystallization phenomenon involving larger segmental motions. Structural changes in the amorphous state of PET during the 'induction period' before the start of crystallization have been detected by X-ray diffraction and might be related to this transition (Imai *et al.*, 1992; Garcia *et al.*, 1999).

   We saw in Section 4.1.1 that the increase of crystallinity with time at constant crystallization temperature follows the Avrami equation (eq. (4.1)). Since in earlier investigations (Santa Cruz *et al.*, 1991) the increase in the H of PET during primary crystallization was shown to be directly proportional to the increase of crystallinity, one may expect that the increase of H will follow a similar relationship as a function of time:

$$H(t) = H_m + (H_M - H_m)\left[1 - \exp(-Gt^n)\right] \tag{4.17}$$

where H_m and H_M represent, respectively, the lowest and highest H values obtained during the crystallization process.

   In Fig. 4.21 $\log\{-\ln[1 - (H(t) - H_m)/(H_M - H_m)]\}$ is plotted as a function of $\log(t)$ for different crystallization temperatures T_c and for samples with and without a nucleating agent. From the evaluation of such curves, one obtains the n and G values directly from the slope and the intercept, respectively, of the straight line. Figure 4.21(*a*) shows data obtained for samples without a nucleating agent. A value of about 3 for the Avrami exponent n is observed for all the temperatures investigated. This value of n is consistent with the concept of three-dimensional growth with a fixed number of nuclei born at $t = 0$. It is known that PET crystallization from the glassy state gives rise to spherulitic materials (Zachmann &

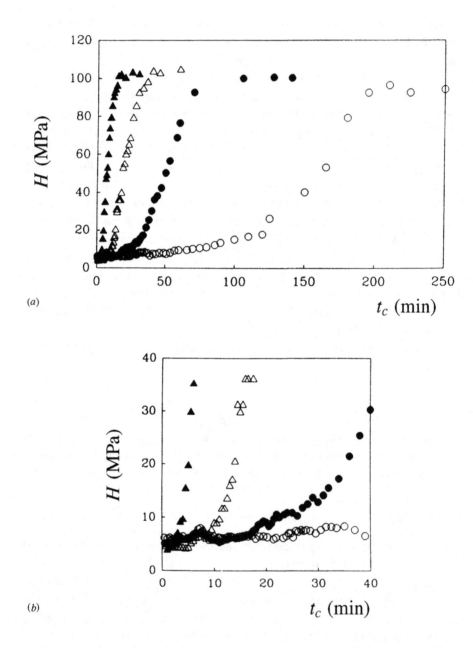

(a)

(b)

Figure 4.20. (a) Variation of microhardness H measured 'in situ' for physically aged PET samples (without nucleating agent) as a function of crystallization time, t_c for different crystallization temperatures, T_c. (b) Microhardness H as a function of crystallization time for the initial stage of crystallization. (▲) $T_c = 117\,°\text{C}$; (△) $T_c = 110\,°\text{C}$; (●) $T_c = 105\,°\text{C}$; (○) $T_c = 100\,°\text{C}$ (From Baltá Calleja *et al.*, 1993.)

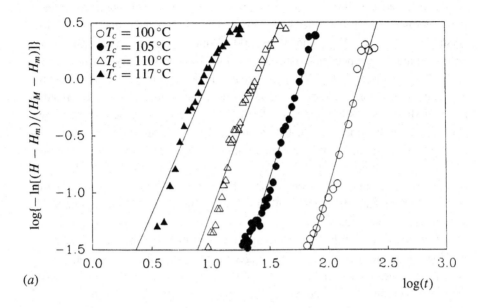

(a)

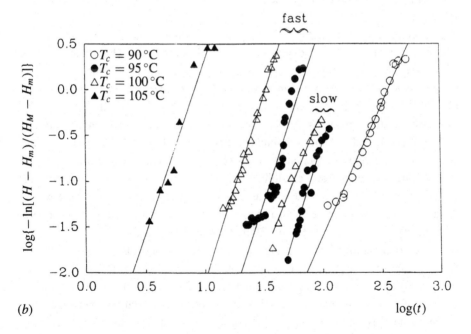

(b)

Figure 4.21. Variation of $\log\{-\ln[1-(H(t)-H_m)/(H_M-H_m)]\}$ as a function of $\log(t)$ for PET samples (a) without (slow) and (b) with (fast) a nucleating agent for different crystallization temperatures T_c. (After Baltá Calleja *et al.*, 1993.)

Stuart, 1960; Groeninckx & Reynaers, 1980). The PET samples used for Fig. 4.21 consist of small spherulites, a few micrometres in size, when observed under a light microscope. On the other hand, for samples with a nucleant (Fig. 4.21(b)) crystallized at $T_c = 95\,°C$ and $T_c = 100\,°C$, one clearly observes two regions corresponding to independent contributions of the fast and slow crystallization rates for each temperature. In Fig. 4.21(b) these two regions correspond to the two pairs of straight lines labelled 'fast' and 'slow'. These two regions can be interpreted in terms of simultaneous crystallization from $t = 0$ due to two independent modes: (1) a fast crystallization mode that starts from the nuclei provided by the nucleating agent which appear at $t = 0$ within the sample. Spherulites would grow from these nuclei up to a limited size at which the nucleating agent is saturated. At this stage, the spherulites do not yet fill the volume of the sample; (2) a slow crystallization mode that starts spontaneously from the existing nuclei and is superposed on the fast crystallization mode.

It is to be noted that the nucleating agent does not change the Avrami exponent n. However, it does provide a much larger number of nuclei (proportional to G) than the number which already exist within the amorphous material. Therefore, the crystallization of a sample with a nucleating agent starts immediately at $t = 0$ and has a faster crystallization rate than the crystallization of a sample without a nucleant.

For temperatures lower than $T_c = 95\,°C$ and higher than $T_c = 100\,°C$ only one slope, with an Avrami exponent $n = 3$, is observed. This means that in these cases the nucleation corresponds to a slow crystallization mode. At low temperatures ($T_c = 90\,°C$) the time required to crystallize is very large ($t_c = 300$ min) because the nuclei appear within the sample very slowly. In this case, the fast crystallization process is not observed because the crystallization temperature is too low. On the other hand, for temperatures above $100\,°C$ ($T_c = 105\,°C$) there is also a single region of crystallization (Fig. 4.21) because the temperature of crystallization is higher, and crystallization proceeds so rapidly (mostly by crystal growth) that it seems reasonable to associate it with the fast mode. Finally, the Avrami constant G increases with T_c indicating that the number of nuclei increases with the crystallization temperature (Baltá Calleja et al., 1993).

The cold crystallization of amorphous films of PET and PEN blends, with different compositions, prepared by coprecipitation from solution followed by melt-pressing for 2 min at $280\,°C$ and quenching in ice-water, has also been followed by measuring the microhardness in real time as a function of crystallization temperature T_c and time t_c (Connor et al., 1997). An analytical model was derived, relating the properties of the individual components to the blend microhardness based on an Avrami-type equation to account for the crystallization of the components upon heating. Fitting of the model to the experimental results revealed a two-step hardening process for the blends. The degree of transesterification of the blends, can be estimated with this model. Heating of the blends above their glass transition rejuvenates the aged material (deageing effect) inducing, as a result, a decrease in the hardness of the PET in the blend.

4.7   Correlation of microhardness to macroscopic mechanical properties

The question of whether microhardness is a property related to the elastic modulus E or the yield stress Y is a problem which has been commented on by Bowman & Bevis (1977). These authors found an experimental relationship between microhardness and modulus and/or yield stress for injection-moulded semicrystalline plastics. According to the classical theory of plasticity the expected microindentation hardness value for a Vickers indenter is approximately equal to three times the yield stress (Tabor's relation). This assumption is only valid for an ideally plastic solid showing sufficiently large deformation with no elastic strains. PE, as we have seen, can be considered to be a two-phase material. Therefore, one might anticipate a certain variation of the $H/Y \sim 3$ ratio depending on the proportion of the compliant to the stiff phase.

In this section results will be discussed in the light of several mechanical models relating the microhardness to yield stress Y and to Young's modulus E.

4.7.1   Mechanical models

According to Tabor, the ratio between the indentation pressure P_m and Y for a Vickers diamond pyramid is about 3.3 ($H = 0.927 P_m$) (Tabor, 1979). This relation applies for materials which behave as ideally plastic solids, but fails for materials in which the elastic strains are non-negligible (March, 1964; Hirst & Howse, 1969).

March (1964) approximated an elastoplastic indentation to the expansion of a spherical cavity under hydrostatic pressure in an infinite elastic-plastic medium. In this simplified model, the indentation pressure P_m is related to the yield stress Y and elastic modulus, E, through:

$$\frac{P_m}{Y} = C + K B \ln Z \tag{4.18}$$

where C and K are constants and B and Z are related to the Y/E ratio through Poisson's ratio, ν. March showed that for a high Y/E ratio (highly elastic materials), the change to a radial flow mode of deformation would occur more easily (low P_m/Y values). Using a Vickers indenter, eq. (4.18) was shown to be valid for a wide variety of materials, ranging from metals to glasses and polymers, leading to best-fit values of $C = 0.28$ and $K = 0.60$. The constants C and K allow for the correction for the possible elastic contraction of the impression on unloading.

Following the expansion cavity theory, Johnson (1985) proposed a model in which the contact surface of the indenter is encased within a hemispherical core.

The hemispherical core of radius a is immediately followed by a plastic zone. The plastic–elastic boundary lies here at a radius c, where $c > a$. The model allows the P_m/Y ratio to be related to a single non-dimensional variable $(E \tan \beta)/Y$, where β is the contact angle between the sample and the indenter ($\beta = 19.7°$ for a Vickers indenter). For $v = 0.5$, Johnson's analysis leads to:

$$\frac{P_m}{Y} = \frac{2}{3}\left[2 + \ln\left(\frac{E \tan \beta}{3Y}\right)\right] \tag{4.19}$$

An additional term in the argument of the logarithm should be introduced for v values other than 0.5. Studman et al. (1977) and Perrot (1977) eliminated certain simplifying assumptions used in Johnson's analysis. These last two approaches led to a modified form of eq. (4.19) which still relates P_m/Y linearly to $\ln[(E \tan \beta)/Y]$ for $v = 0.5$ but which requires new values for the numerical factors.

4.7.2   Correlation between hardness and yield stress

Figure 4.22 shows a plot of H as a function of the tensile yield stress Y_t (solid circles) and the compressive yield stress Y_c (open circles) for a variety of PE

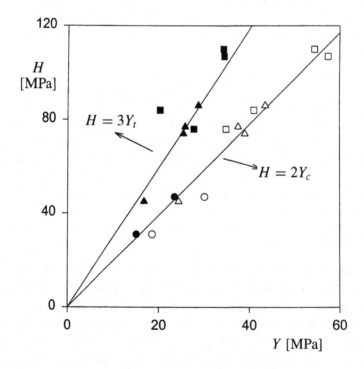

Figure 4.22. Variation of hardness H with tensile Y_t (solid symbols) and compressive Y_c yield stresses (open symbols). $\bigcirc$, $\bullet$, compression-moulded samples; $\triangle$, $\blacktriangle$, annealed samples at atmospheric pressure; $\square$, $\blacksquare$, CEPE samples. (From Flores et al., 2000.)

samples with different morphologies (Flores *et al.*, 2000). In both cases, the higher H and Y values correspond to chain-extended samples while the lower ones are associated with the starting chain-folded material (see Flores *et al.*, 1999, for a detailed discussion). It is seen that in all cases Y_c is larger than Y_t. This difference has been ascribed to the effect of the hydrostatic component on stress on isotropic polymers including PE (Ward, 1971). The data in Fig. 4.22 fit onto two straight lines which pass through the origin yielding $H/Y_t \sim 3$ and $H/Y_c \sim 2$. Earlier studies of melt-crystallized PE had already indicated that $H \sim 3Y_t$ when the strain rate in the tensile tests is comparable to that employed in the microhardness tests (Baltá Calleja *et al.*, 1995). A closer inspection of Fig. 4.22 reveals slight deviations from $H/Y_t \sim 3$, the original moulded samples exhibiting values $H/Y_t \sim 2$ while the chain-extended samples show $H/Y_t \sim 3$–4. On the other hand, the H/Y_c ratio deviates considerably from that predicted by the theory of plasticity ($H/Y_c \sim 3$) even for highly crystalline samples such as CEPE, where one would expect the elastic strains to be minimized.

4.7.3 Comparison with elastoplastic models

The above results evidence the elastoplastic response to indentation for the PE samples discussed. Thus, it is convenient to discuss the microhardness–yield stress correlation in the light of models that account for the elastic–plastic strains under a compressive stress. Figure 4.23 shows the semilogarithmic plot of the H/Y_c vs $(E_c \tan \beta)/Y_c$ data for all the PE samples investigated. The E_c/Y_c ratio is a measure of the reciprocal value of the elastic strain of the material. Hence, the observed tendency of H/Y_c to increase towards $H \sim 3Y_c$ (ideal plastic behaviour) can be related to a decrease of elastic strains. The H/Y_c ratio is lowest for the compression-moulded (chain-folded) samples, where the fraction of the compliant (amorphous) phase is the highest. The various straight lines plotted in Fig. 4.23 are drawn according to the different models of elastoplastic indentation (using $H = 0.927 P_m$). In the present calculations, the approximation $\nu = 0.5$ was used. The theoretical straight lines in Fig. 4.23 are represented within the region of elastoplastic contact. This approximately comprises H/Y_c values from 0.5 (the first yield for a Vickers indenter occurs at $P_m \sim 0.5Y$) up to 3 (fully plastic deformation). For $(E_c \tan \beta)/Y_c$ values higher than 40–60, the rigid-plastic mode of deformation prevails while for $(E_c \tan \beta)/Y_c < 1$, the response is largely elastic. Data for some work-hardened metals have been also included to illustrate the H vs Y_c correlation for fully plastic materials, $H = 3Y_c$ (horizontal dashed-dotted line) in Fig. 4.23.

The theoretical straight lines in Fig. 4.23 are in consonance with the experimental data for all the PE samples. The best fit possibly corresponds to Johnson's model (1985), although the difference with other approaches is small within the range of H/Y_c values considered.

According to Johnson's model, the P_m/E_c ratio, where E_c is the modulus in compression, is described by:

$$\frac{P_m}{E_c} = \frac{Y_c}{E_c}\frac{2}{3}\left[2 + \ln\left(\frac{E_c \tan \beta}{e Y_c}\right)\right] \qquad (4.20)$$

Analogous equations can be derived for the mechanical models of March (1964), Studman *et al.* (1977) and Perrot (1977). Figure 4.24 illustrates the curve (solid line) calculated according to eq. (4.20) relating P_m/E_c with Y_c/E_c and using a value of $H = 0.927 P_m$. Data for work-hardened metals are included together with the straight line indicating fully plastic behaviour (dashed line). For $Y_c/E_c < 0.01[(E_c \tan \beta)/Y_c > 40]$, the material behaves as ideally plastic and Tabor's relation applies ($H/Y_c \sim 3$). The data plotted in Fig. 4.24 for all the PE samples are shown to be in good agreement with eq. (4.18). As the Y_c/E_c ratio diminishes towards the fully plastic deformation, the H/E_c values decrease. CEPE shows the lowest Y_c/E_c values in agreement with the higher crystallinity values found for this material.

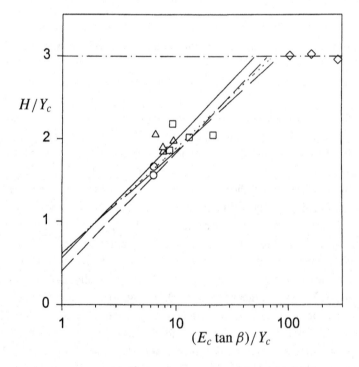

Figure 4.23. Plot of H/Y_c vs $\log[(\tan \beta E_c)/Y_c]$ for: $\circ$, compression-moulded PE; $\triangle$, annealed PE at atmospheric pressure; $\square$, CEPE; $\diamond$, work-hardened metals. The straight lines follow the theoretical models of elastoplastic indentation from: $(\cdots)$ Marsh; $(\textemdash)$ Johnson; (- - -) Studman *et al.*; $(\textemdash\,\textemdash)$ Perrot. (After Flores *et al.*, 2000.)

4.7.4   Correlation between hardness and elastic modulus

Figure 4.25 illustrates the plot of H vs E_t, where E_t is the modulus in tension for PE samples (Flores *et al.*, 2000), including earlier data from Santa Cruz (1991). Micro-hardness shows a general tendency to increase with increasing elastic modulus.

In the previous section we found $H \sim 3Y_t$. Moreover, Struik (1991) developed a model that takes account of the intermolecular forces between adjacent molecules and relates the yield stress to the elastic modulus E which predicts:

$$Y \approx E/30 \tag{4.21}$$

Struik successfully tested eq. (4.21) for various semicrystalline and amorphous polymers subjected to tensile experiments. If we make use of eq. (4.21) together with $H = 3Y_t$, then we obtain $H \approx E/10$. This correlation has been represented in Fig. 4.25 by a straight line. The experimental data corresponding to chain-folded PE are in fair agreement with $H = E_t/10$. However, data for CEPE, most especially for high molecular weight CEPE ($E = 1620$ MPa), show severe deviations from the theoretical predictions. This result seems to imply that here lamellar thickness is no longer the dominant morphological factor. Instead in chain-extended material, lamellar connectedness plays a more decisive role in determining E. In fact, the

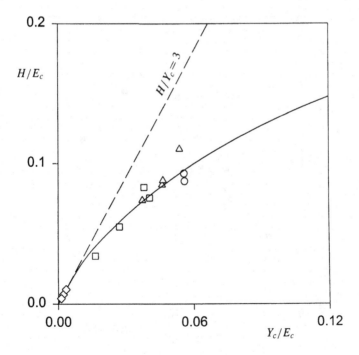

Figure 4.24. Plot of H/E_c as a function of Y_c/E_c, where E_c is the modulus in compression, for the PE samples investigated. Symbols are as in Fig. 4.23. The theoretical curve (continuous line) is drawn according to Johnson's model. (From Flores *et al.*, 2000.)

main reason for ductility in ultra-stiff CEPE has to be sought in the strength and toughness of the interlamellar material, which directly increases with the number of interlamellar connections (Attenburrow & Bassett, 1977).

The following conclusions can be drawn from the foregoing section:

- For melt-crystallized and chain-extended PE with a wide range of morphologies the relationship $H \sim 3Y_t$ applies, provided the strain rates employed in the tensile and indentation tests are comparable.

- For the yield stress in compression, deviations from Tabor's relation giving values of $H \sim 2Y_c$ are found. This is presumably due to the elastic strain of the indented material. A detailed analysis of the H/Y_c ratio on the basis of mechanical models of elastoplastic indentation reveals that H/Y_c linearly increases with $\ln[(\tan \beta E_c)/Y_c]$. Compression-moulded (chain-folded) PE samples, which present the lowest crystallinity of all the samples investigated, also show the lowest H/Y_c ratio as a consequence of the comparatively large elastic strains.

- The H/E_c ratio is shown to decrease with decreasing Y_c/E_c ratio following the theoretical prediction for elastoplastic indentation. CEPE shows the

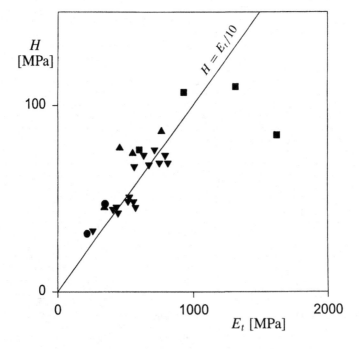

Figure 4.25. Microhardness H variation with E_t, modulus in tension, for: ●, compression-moulded PE; ▲, annealed PE at atmospheric pressure; ■, CEPE; ▼, melt-crystallized (chain-folded) PE as taken. (From Flores *et al.*, 2000.)

lowest H/E_c values in agreement with an enhanced plastic behaviour in contrast to chain-folded PE.

■ Microhardness is related to Young's modulus, as derived from tensile experiments, through $H \sim E_t/10$, in agreement with Struik's predictions of $Y \sim E/30$.

4.8 References

Ania, F., Kilian, H.G. & Baltá Calleja, F.J. (1986) *J. Mater. Sci.* **5**, 1183.

Arlie, J.P., Spegt, P. & Skoulios, A. (1966) *Makromol. Chem.* **99**, 160.

Arlie, J.P., Spegt, P. & Skoulios, A. (1967) *Makromol. Chem.* **104**, 212.

Asano, T. & Fujiwara, Y. (1978) *Polymer*, **19**, 99.

Attenburrow, G.E. & Bassett, D.C. (1977) *J. Mater Sci.* **12**, 192.

Avrami, M.J. (1939) *J. Chem. Phys.* **7**, 1103.

Baltá Calleja, F.J. (1976) *Colloid Polym. Sci.* **254**, 258.

Baltá Calleja, F.J. (1985) *Adv. Polym. Sci.* **66**, 117.

Baltá Calleja, F.J. (1994) *Trends in Polym. Sci.* **2**, 419.

Baltá Calleja, F.J. & Kilian, H.G. (1985) *Colloid Polym. Sci.* **263**, 697.

Baltá Calleja, F.J. & Santa Cruz, C. (1996) *Acta Polym.* **47**, 303.

Baltá Calleja, F.J. & Vonk, C.G. (1989) *X-ray Scattering of Synthetic Polymers*, Elsevier, Amsterdam.

Baltá Calleja, F.J., Giri, L., Michler, G.H. & Naumann, I. (1997) *Polymer* **38**, 5769.

Baltá Calleja, F.J., Giri, L., Ward, I.M. & Cansfield, D.L.M. (1995) *J. Mater. Sci.* **30**, 1139.

Baltá Calleja, F.J., Hosemann, R. & Wilke, W. (1969) *J. Polym. Sci.: Part C* **16**, 4329.

Baltá Calleja, F.J., Martínez-Salazar, J. & Asano, T. (1988) *J. Mater. Sci. Lett.* **7**, 165.

Baltá Calleja, F.J., Martínez-Salazar, J., Cackovic, H. & Loboda-Cackovic, J. (1981) *J. Mater. Sci.* **16**, 739.

Baltá Calleja, F.J., Öhm, O & Bayer, R.K. (1994) *Polymer* **35**, 4775.

Baltá Calleja, F.J., Santa Cruz, C. & Asano, T. (1993) *J. Polym. Sci. Polym. Phys. Ed.* **31**, 557.

Baltá Calleja, F.J., Santa Cruz, C., Asano, T. & Sawatari, C. (1990a) *Macromolecules* **23**, 5352.

Baltá Calleja, F.J., Santa Cruz, C., Bayer, R.K. & Kilian, H.G. (1990b) *Colloid Polym. Sci.* **268**, 440.

Baltá Calleja, F.J., Santa Cruz, C., González Arche, A. & López Cabarcos, E. (1992) *J. Mater. Sci.* **27**, 2124.

Bassett, D.C. (1981) *Principles of Polymer Morphology*, Cambridge University Press, Cambridge.

Bassett, D.C. & Carder, D.R. (1973) *Phil. Mag.* **28**, 513.

Bayer, R.K., Baltá Calleja, F.J. & Kilian, H.G. (1997) *Colloid Polym. Sci.* **275**, 432.

Bowmann, J. & Bevis, M. (1977) *Colloid Polym. Sci.* **255**, 954.

Buckley, C.P. & Kovacs, A.J. (1976) *Colloid Polym. Sci.* **254**, 695.

Connor, M.T., García Gutiérrez, M.C., Rueda, D.R. & Baltá Calleja, F.J. (1997) *J. Mater. Sci.* **32**, 5615.

Crist, B., Fisher, C.J. & Howard, P.R. (1989) *Macromolecules* **22**, 1709.

Darlix, B., Monasse, B. & Montmitonnet, P. (1986) *Polym. Test* **6**, 107.

Deslandes, Y., Alva Rosa, E., Brisse, F. & Meneghini, T. (1991) *J. Mater. Sci.* **26**, 2769.

Dunn, P.E. & Carr, S.H. (1989) *MRS Bulletin* **14**(2), 22.

Eisenriegler, E. (1993) *Polymers Near Surfaces*, World Scientific, Singapore.

Evans, U.R. (1945) *Trans. Faraday Soc.* **14**, 365.

Fakirov, S. (1990) in *Solid State Behaviour of Linear Polyesters and Polyamides* (Schultz, J.M. & Fakirov, S., eds.) Prentice Hall, Englewood Clifs, New Jersey, p. 1.

Flores, A., Aurrekoetxea, J., Gensler, R., Kausch, H.H. & Baltá Calleja, F.J. (1998) *Colloid Polym. Sci.* **276**, 786.

Flores A., Baltá Calleja, F.J., Attenburrow, G.E. & Bassett, D.C. (2000) *Polymer* **41**, 5431.

Flores, A., Baltá Calleja, F.J. & Bassett, D.C. (1999) *J. Polym. Sci., Polym. Phys. Ed* **37**, 3151.

Fujiwara, Y. (1968) *Kolloid 2.2. Polymere* **226**, 135.

Garcia, M.C., Rueda, D.R. & Baltá Calleja, F.J. (1999) *Polym. J.* **31**, 806.

Gedde, U.W. (1995) *Polymer Physics*, Chapman Hall, London.

Groeninckx, G. & Reynaers, H. (1980) *J. Polym. Sci. Polym. Phys. Ed.* **B18**, 1325.

Grubb, D.T. & Keller, A. (1980) *J. Polym. Sci. Polym. Phys. Ed.* **18**, 207.

Hirami, M., Baltá Calleja, F.J. & Flores, A. (1999) *Polym. J.* **31**, 747.

Hirst, W. & Howse, M.G.J.W. (1969) *Proc. Roy. Soc. London* **311**, 429.

Hosemann, R., Wilke, W. & Baltá Calleja, F.J. (1966) *Acta Cryst.* **21**, 118.

Imai, M., Mori, K., Mizukami, T., Kaji, K. & Kanaya, T. (1992) *Polymer* **33**, 4451.

Ingram, P. & Peterlin, A. (1968) in *Encyclopedia of Polymer Science and Technology*, Vol. 9, John Wiley & Sons, New York, p. 203.

Johnson, K.L. (1985) *Contact Mechanics*, Cambridge University Press., Cambridge.

Kanig, G. (1975) *Prog. Colloid & Polym. Sci.* **57**, 176.

Kausch, H.H. & Jud, K. (1982) *Plastic and Rubber Processing and Applications* **2**, 265.

Keller, A. (1959) *J. Polym. Sci.* **39**, 151.

Keller, A. (1968) *Rep. Prog. Phys.* **31**, 632.

Koga, K. & Ohigashi, H. (1986) *J. Appl. Phys.* **59**, 2142.

Krevelen, D.W. van (1976) *Properties of Polymers*, Elsevier, Amsterdam.

Krumova, M., Fakirov, S., Baltá Calleja, F.J. & Evstatiev, M. (1998) *J. Mater. Sci.* **33**, 2857.

Lawn, B.R. & Howes, V.R. (1981) *J. Mater. Sci.* **16**, 2745.

Lovinger, A.J. (1985) *Macromolecules* **18**, 910.

Lovinger, A.J., Davies, G.T., Furukawa, T. & Broadhurst, M.G. (1982) *Macromolecules* **15**, 323.

Lovinger, A.J., Furukawa, T., Davies, G.T. & Broadhurst, M.G. (1983) *Polymer* **24**, 1225.

Mandelkern, L. (1979) *Faraday Discuss. Chem. Soc.* **68**, 310.

Marsh, D.M. (1964) *Proc. Phys. Soc. London* **A279**, 420.

Martínez-Salazar, J., García Tijero, J.M. & Baltá Calleja, F.J. (1988) *J. Mater. Sci.* **23**, 862.

Nielsen, L.E. (1954) *J. Appl. Phys.* **25**, 1209.

Perrot, C.M. (1977) *Wear* **45**, 293.

Peterlin, A. (1971) *J. Mater. Sci.* **6**, 490.

Peterlin, A. (1973) *J. Macromol. Sci. Phys.* **B8**, 83.

Peterlin, A. (1987) *Colloid Polym. Sci.* **265**, 357.

Popli, R. & Mandelkern, L. (1987) *J. Polym. Sci., Polym. Phys. Ed.* **25**, 441.

Reding, F.P. (1958) *J. Polym. Sci.* **32**, 487.

Santa Cruz, C. (1991) PhD Thesis, Universidad Autónoma de Madrid.

Santa Cruz, C., Baltá Calleja, F.J., Asano, T. & Ward, I.M. (1993) *Phil. Mag.* **A68**, 209.

Santa Cruz, C., Baltá Calleja, F.J., Zachmann, H.G., Stribeck, N. & Asano, T. (1991) *J. Polym. Sci., Polym. Phys. Ed.* **29**, 819.

Struik, L.C.E. (1991) *J. Noncryst. Solids* **131–133**, 395.

Studman, C.J., Moore, M.A. & Jones, S.E. (1977) *J. Phys. D: Appl. Phys.* **10**, 949.

Tabor, D. (1979) *Gases, Liquids and Solids*, second edition, Cambridge University Press, Cambridge, p. 188.

Tashiro, K. & Kobayashi, M. (1985) *Polymer* **27**, 67.

Tashiro, K. & Tadokoro, H. (1987) in *Encyclopedia of Polymer Science and Engineering*, Supplement, John Wiley & Sons, New York, p. 187.

Tashiro, K., Nakamura, M., Kobayashi, M., Chatani, Y. & Tadokoro, H. (1984) *Macromolecules* **17**, 1452.

Tashiro, K., Takano, K., Kobayashi, M., Chatani, Y. & Tadokoro, H. (1981) *Polymer* **22**, 1312.

Vanderdonckt, C., Krumova, M., Baltá Calleja, F.J., Zachmann, H.G. & Fakirov, S. (1998) *Colloid Polym. Sci.* **276**, 138.

Ward, I.M. (1971) *J. Polym. Sci.* **C32**, 195.

Ward, I.M. (1982) *Developments in Oriented Polymers*–1, Applied Science Publishers, London, Chapter 6.

Ward, I.M. & Hadley, D.W. (1993) *An Introduction to the Mechanical Properties of Solid Polymers*, John Wiley & Sons, Chichester, p. 221.

Williamson, G.R., Wright, B. & Howard, R.N. (1964) *J. Appl. Chem.* **14**, 131.

Wunderlich, B. (1976) *Macromolecular Physics*, Vol. 2, Academic Press, New York, p. 131.

Young, R.J. (1988) *Mater. Forum* **22**, 210.

Zachmann, H.G. & Stuart, H.A. (1960) *Makromol. Chem.* **49**, 131.

Chapter 5

Microhardness of polymer blends, copolymers and composites

5.1   Blends of polyolefins

5.1.1   Model predictions in PE blends

The interest in multicomponent materials, in the past, has led to many attempts to relate their mechanical behaviour to that of the constituent phases (Hull, 1981). Several theoretical developments have concentrated on the study of the elastic moduli of two-component systems (Arridge, 1975; Peterlin, 1973). Specifically, the application of composite theories to relationships between elastic modulus and microstructure applies for semicrystalline polymers exhibiting distinct crystalline and amorphous phases (Andrews, 1974). Furthermore, as discussed in Chapter 4, the elastic modulus has been shown to be correlated to microhardness for lamellar PE. In addition, H has been shown to be a property that describes a semicrystalline polymer as a composite material consisting of stiff (crystals) and soft, compliant elements. Application of this concept to lamellar PE involves, however, certain difficulties. This material has a microstructure that requires specific methods of analysis involving the calculation of the volume fraction of crystallized material, crystal shape and dimensions, etc. (Baltá Calleja *et al.*, 1981).

For this reason, it is of interest to investigate the H of a model system composed of varying mixtures of two types of PE with well-differentiated microhardness values in such a way that the experimental microhardness data derived can be compared with predictions for the various component arrangements. In addition, the measurement of H of these blends at high temperature can provide direct informa-tion on the microhardness of the disordered phase. The mechanical characterization

of these blends is also of interest from the viewpoint of the production of materials
with novel properties.

As a first example we will discuss the case of blends of high-density (HD) and
low-density (LD) ($3CH_3/100CH_2$) PE with $M_w \sim 50 \times 10^3$, prepared in a wide
range of compositions (Martínez-Salazar & Baltá Calleja, 1985). The samples were
crystallized from the melt by quenching to room temperature. The intimate mixture
of the two molecular species was indicated by the decrease in the crystallization
temperature of the linear polymer component with increasing proportion of the
branched polymer species. In addition, X-ray scattering analysis revealed that
the branched polymer molecules are not incorporated within the linear polymer
crystals. This result is also confirmed by the presence of two melting peaks in the
thermograms of the blends.

Figure 5.1. shows the linear increase of the microhardness of the low-density
polymer with increasing concentration of the linear material, from 20 MPa up to
about 70 MPa , for 100% of the HD component. A similar linear increase is obtained
for the slowly crystallized materials (see arrows in Fig. 5.1.). The H values are,
however, larger owing to the larger crystal thicknesses of both components (Baltá
Calleja, 1976). Using the additivity law for a system comprising two types of
crystals, LD and HD, then:

$$H_{blend} = H_1 w + H_2(1 - w) \tag{5.1}$$

which is identified as a parallel composite model, where H_1 and H_2 are the microhardnesses of the two independent components and w is the weight concentration of the HDPE. Equation (5.1) emphasizes the validity of a distinct phase
microhardness for these PE blends. This is probably due to the above-mentioned
molecular segregation at a crystal level. The microhardness for the blends measured
at high temperature is substantially lower. This is a consequence of the decrease
in the chain packing within the crystals (see Section 2.6). The decrease in H
for the quenched samples measured at high temperature (110 °C) is also linear,
however, with a lower slope. This is due to the fact that the rate of decrease of
the microhardness is proportional to the degree of crystallinity. Hence, the HD
component contributes to a faster decrease in H than does the LD component. At
the temperature at which the LD component melts (100 °C) the microhardness for
100% LDPE obviously cannot be determined experimentally. However, it can be
derived by extrapolation. The extrapolation (see Fig. 5.1) yields a value for the
microhardness of amorphous PE of approximately 0.5 MPa. Such a value is in
agreement with estimated H data from melt-crystallized PE samples extrapolated
to the density of amorphous PE at room temperature (Baltá Calleja, 1985). In
summary, the results described confirm that the microhardness of the studied PE
blends can be explained in terms of a simple additive system of two independent
components H_1 and H_2.

These studies have been extended to the case of blends based on recycled PE wastes. Recycled grades of HDPE are being incorporated as new materials for packaging in the plastics industry (Ehring, 1992; Gibbs, 1990). Extrusion blow-moulding is one of the techniques commonly used to produce recycled PE films and bottles. Blends based on waste PE with different molecular-weight grades may be employed in the preparation of extruded blown films for different applications. To study the microhardness and other properties of these materials, two samples of LDPE waste (agricultural and packaging films) with number-average molecular weights, M_n, of 15 500 ($A1$) and 22 700 ($A2$) and a HDPE waste (industrial scrap) of $M_n = 38\,600$ (B) were selected to prepare blends $A1/B$ and $A2/B$ in a wide range of compositions. The blends were prepared by mixing the two components on a two-roll mill at 145–150 °C for 10 min. The milled material was hot-pressed at 140 °C for 3 min to prepare films of about 0.3 mm thickness (Rueda *et al.*, 1994).

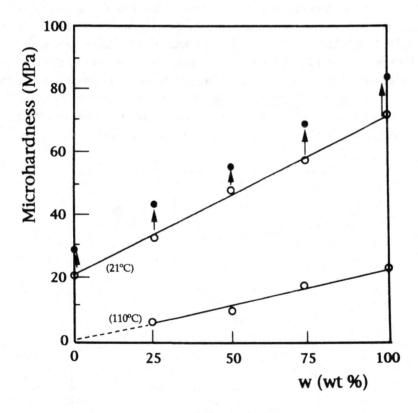

Figure 5.1. Microhardness (H) of LDPE and HDPE samples quenched from the melt as a function of the concentration of the latter component. H data taken at 21 and 110 °C are shown (○). Results from slowly cooled samples are also given (●). (From Martínez-Salazar & Baltá Calleja, 1985.)

Figure 5.2 shows the linear relationship found between the microhardness and the yield stress for the two series of blends $A1/B$ and $A2/B$. However, as we will discuss later, the slope of this plot differs from that found for other PE samples (Baltá Calleja, 1985). The simultaneous increase of both H and tensile yield stress Y_t with increasing content of the HDPE component results in a linear correlation between these two mechanical properties with a ratio $H/Y_t \simeq 2$ (Fig. 5.2). This ratio is significantly smaller than that previously found for PE ($H \simeq 3Y_t$) (see Section 4.7.3). The smaller H/Y_t ratio found for these materials can be related to the relatively high deformation speed (50 mm min^{-1}) used in a macroscopic measurement, which is about 40 times higher than that used in a microhardness test (see Baltá Calleja et al., 1995).

Figure 5.3 illustrates the variation of H for $A1/B$ blends with increasing content of the high-molecular-weight component, w_B, indicating a deviation from an additive behaviour of H for the two independent components. In the case of the $A2/B$ blends a smaller deviation of H from the additivity of H was found. The deviation of the microhardness from a linear behaviour in the H vs composition plot (Fig. 5.3) can be explained by examining the thermograms of the blends (Rueda et al., 1994). It is known that the endothermic area of DSC thermograms represents the fraction of crystallized material in the sample. The presence in the thermograms of a high-temperature peak at $\sim$139 °C and other low-temperature peaks in the range 110–120 °C suggests the coexistence of two types of lamellae: thick lamellae (15 nm) and thinner ones (7.0–9.5 nm). The former correspond essentially to the B component while the latter are a mixture of those from the A component

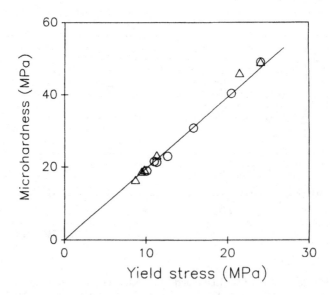

Figure 5.2. Correlation between microhardness and yield stress for (O) A1/B and (△) A2/B blends from recycled PE wastes (Rueda et al., 1994) (see the text for details).

plus additional intermediate ones which arise from a new endothermic peak, or shoulder, that appears in the range 115–120 °C. The fraction of material (from the B component) associated with the high-temperature peak (HTP) can be calculated using the expression:

$$w'_B(\%) = \left(\frac{\text{Area of HTP}}{\text{Total area}} \right) \left(\frac{\delta H_{sample}}{\delta H_B} \right) \times 100$$

where δH is the heat of fusion of the samples and δH_B is the heat of fusion from component B. From the remaining area of the thermogram which is associated with the fusion of crystals of low-density material, the fraction of material A (w'_A) can be calculated. The fractions w'_A and w'_B can then be represented as a function of the nominal weight percentage of B for $A1/B$ samples (Rueda et al., 1994). A clear deviation from the bisector line is observed for compositions $w_B > 10\%$. For all the samples investigated $w'_A + w'_B \simeq 1$.

We can now use the new composition values w'_A and w'_B to derive the microhardness H for blends of recycled PE in terms of a composite comprising two populations of crystals (the thick ones and the thinner ones). According to the additivity law

$$H_{cal} = H_B w'_B + H_A w'_A \tag{5.2}$$

where H_B and H_A are the microhardnesses of the individual B and A components.

The good agreement obtained between calculated and experimental data (Rueda et al., 1994) confirms that the microhardness of these blends, if one uses the new

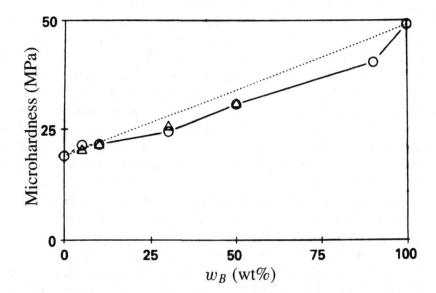

Figure 5.3. Microhardness *vs* composition w_B for A1/B samples: (○) measured hardness values; (△) Calculated values (Rueda *et al.*, 1994).

composition data derived from calorimetry, can be explained again in terms of a simple additive system of the two single components H_B and H_A. The reason for the deviation of H (and also Y_t) from linear additive behaviour with nominal composition can be attributed to a great reorganization of molecules forming the hard crystalline lamellae during the blending process. As pointed out above, new crystals are formed after blending as revealed by the presence of the intermediate endothermic peak in the DSC traces. This intermediate peak most probably arises from a molecular segregation and recombination of both more defective (from component B) and less defective (from component A) molecules during crystallization. As we have seen in Chapter 4 the microhardness of PE crystals is very dependent on crystal size and perfection, factors which also influence the melting temperature.

5.1.2   Deviations from the microhardness additivity law: PE/PP blends

The preceding studies on blends of PE with various density values have been complemented by others on blends of PE with PP over a wide composition range (Martínez-Salazar et al., 1988). It was found that the microhardness of PE/PP blends can also be described in terms of a parallel additive system of two independent components H_{PE} and H_{PP}. However, treatments, such as (a) crystallization of the PP phase in the presence of molten PE within the blends, and (b) annealing of the PE phase, leaving the PP component unaltered, induce deviations from the microhardness additive behaviour of the independent components. The presence of PE throughout the range of blends inhibits the crystallization of PP, inducing a depression of crystallinity which, in turn, causes a depression of the microhardness expected from the additivity law. Conversely, the presence of PP in the blends substantially modifies the annealing behaviour of the PE crystals (fraction of initial lamellae annealed) inducing a similar depression from the H additivity values.

Sawatari et al. (1987) suggested that the easy drawability of PE/i-PP blend films prepared by gel crystallization from semidilute solutions, yielding draw ratios larger than 50, is due to the existence of a relatively low level of molecular entanglements between the PE and i-PP lamellar crystals. Thus, PE/i-PP blend gel films may offer an interesting model system with presumably a low density of surface defects, which can be tested by the microhardness technique. In relation to this contention it is interesting to study the microhardness of ultra-high-molecular-weight PE/i-PP blend films and examine whether the microhardness-derived constant b which is sensitive to surface effects (see eq. (4.14)) depends on PE/i-PP composition.

PE/i-PP blend films were prepared by gel crystallization from semidilute decalin solution as reported by Baltá Calleja et al. (1990b), using ultra-high-molecular-weight PE ($M_2 = 6 \times 10^6$) and i-PP ($M_w = 4.4 \times 10^6$). In addition to the individual PE and i-PP homopolymer dry gels, Baltá Calleja et al. investigated PE/i-PP compositions of 75/25, 50/50 and 25/75. For all compositions a concentration of about

0.45 g/100 ml was shown to be most convenient to ensure the maximum draw ratio of each blend film.

Figure 5.4 illustrates the experimentally obtained variation of H as a function of w, the weight concentration of the PE component (curve 3). It is seen that the H values for the initial PE and i-PP gel films ($H^{PE} = 105$ MPa and $H^{PP} = 116$ MPa) do not differ substantially from each other. In spite of the fact that the gel films (especially the homopolymers) show a porous microstructure, the H values are much larger ($\sim$50%) than those obtained for conventionally crystallized samples

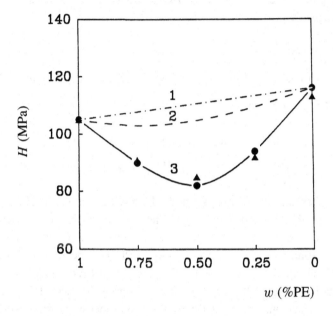

Figure 5.4. Microhardness H of PE/i-PP blended gel films as a function of PE concentration w; (1) additivity behaviour from eq. (5.3) using the w_c values of the individual homopolymers; (2) H values using w_c^{PE} and w_c^{PP} data (Table 5.1). (3) ($\bullet$) experimental data; ($\triangle$) calculated values from Table 5.2. (From Baltá Calleja *et al.*, 1990a.)

Table 5.1. *Experimental values of microhardness H and crystallinity of PE w_c^{PE} and i-PP w_c^{PP} blended gel films as a function of composition (by weight).*

PE/i-PP	H (MPa)	w_c^{PE}	w_c^{PP}
100/0	105	0.80	
75/25	90	0.77	0.39
50/50	82	0.74	0.43
25/75	94	0.71	0.45
0/100	116		0.49

(Baltá Calleja, 1985). This implies that the microhardness in these materials is not substantially affected by the presence of microvoids a few micrometres in diameter (Matsuo *et al.*, 1984). Furthermore, a very conspicuous deviation from the additivity law (straight line 1):

$$H = wH^{PE} + (1 - w)H^{PP} \tag{5.3}$$

with increasing PE concentration w is detected. Since the glass transition temperature for PE is much lower than room temperature, the microhardness contribution of the amorphous phase is $H_a^{PE} \simeq 0$. Hence, for PE using the additivity model of eq. (5.1) we may write

$$H^{PE} = w_c^{PE}\, H_c^{PE} \tag{5.4}$$

On the other hand, since for i-PP $H_a^{PP} \neq 0$ (see Section 4.2.2), for i-PP

$$H^{PP} = w_c^{PP}\, H_c^{PP} + (1 - w_c^{PP})H_a^{PP} \tag{5.5}$$

By combining eqs. (5.4)–(5.6) we are led to the expression

$$H = w\, w_c^{PE}\, H_c^{PE} + (1 - w)w_c^{PP}\, H_c^{PP} + (1 - w)(1 - w_c^{PP})H_a^{PP} \tag{5.6}$$

which describes the microhardness of the blended gel films in terms of the microhardnesses of the independent crystalline and amorphous components. From Table 5.1 one sees that the i-PP component hampers the crystallization ability of PE and, therefore, the degree of crystallinity of PE w_c^{PE} decreases with increasing i-PP concentration. On the other hand, the presence of PE crystals also inhibits the crystallization level of i-PP and w_c^{PP} likewise diminishes with PE concentration. If we take into account the crystallinity depression measured for the PE and i-PP components in eq. (5.7), use $H_c^{PE} = 130$ MPa and $H_c^{PP} = 145$ MPa, and let $H_a^{PP} = 90$ MPa, we are then led to curve 2 in Fig. 5.4. In another investigation it has been shown that a decrease in crystallinity of PE/i-PP blends crystallized from the melt also led to a clear deviation of H from the additivity law (Martínez-Salazar

Table 5.2. *Calculated data, using eqs. (5.3)–(5.6), as a function of composition w (see text).*

PE/i-PP	b_{PE} (Å)	b_{PP} (Å)	H_c^{PE} (MPa)	H_c^{PP} (MPa)	H^{cal} (MPa)
100/0	28		131		105
75/25	40	35	118	81	90
50/50	57	30	103	90	83
25/75	61	14	99	116	113
0/100		5		137	113

et al., 1988). However, in the case of the present blended gel films the crystallinity depression cannot account for the substantial deviation of the experimental H values from the additivity law (Fig. 5.4). Hence, according to eq. (5.6), the only possibility left to explain the conspicuous microhardness depression found is that the crystal microhardnesses, H_c^{PE} and H_c^{PP}, themselves decrease, as a consequence of the coexistence of PE and *i*-PP components. At this point, one may enquire what the origin for such H_c^{PE} and H_c^{PP} depression can be. If one uses eq. (4.8), then one immediately sees that, for crystals with given thicknesses ℓ_c ($\ell_c^{PE} = w_c^{PE}L_{PE}$ and $\ell_c^{PP} = w_c^{PP}L_{PP}$, where $L_{PE} = 119$ Å and $L_{PP} = 105$ Å are the X-ray long periods of the individual homopolymers), H_c depends only on the parameter $b = 2\sigma_e/\Delta h$, which furnishes a measure of the surface free energy of the laminar crystals. The latter depends on the level of defects located on the surface boundary of the crystals.

If we use for PE a value of $H_c^0 = 170$ MPa (Baltá Calleja & Kilian, 1985) and for *i*-PP a value of $H_c^0 = 150$ MPa (Baltá Calleja & Martínez-Salazar, 1988) by adjusting the b value in eq. (4.8) for PE and *i*-PP, respectively, we can obtain from eq. (5.6) the H_c values for the different compositions which give the best fit between experimental and calculated H values (see Table 5.2). The b values for PE and *i*-PP

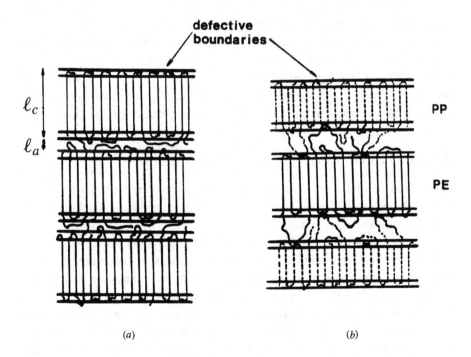

(*a*)   (*b*)

Figure 5.5. Schematics of dry gel lamellae for: (*a*) PE with chain folds and loops at the surface; (*b*) PE/*i*-PP blend showing PP entanglements on the defective surface of PE lamellae (Baltá Calleja *et al.*, 1990a.)

that give the best fit between H-calculated and H-experimental in Fig. 5.4 are listed in Table 5.2. These data show that: firstly, the b_{PE} parameter for blended gel films of PE and for the individual PE homopolymer is 4–9 times smaller than that for conventionally melt-crystallized PE (Baltá Calleja et al., 1990b); and secondly, the b_{PE} and b_{PP} values for the blended gel films gradually increase with the content of i-PP and PE crystals, respectively, in the films. These results could tentatively be explained in light of the findings of Sawatari et al. (1987) as due to an increase in the number of molecular entanglements between PE and the neighbouring i-PP crystals with increasing i-PP content up to 75%.

This is consistent with the view that the PE homopolymer gel lamellae are similar structures to solution-crystallized chain-folded PE with almost no entanglements (Fig. 5.5(a)). However, with increasing i-PP concentration the increase in σ_e from 41 up to 54 erg cm^{-2} indicates that the defect excess energy term, σ_e, gradually increases up to values of about 16.4 erg cm^{-2}. One could speculate that this result is in accordance with an increase in the number of entanglements formed by neighbouring i-PP chains on the surface of PE lamellae (Fig. 5.5(b)). It is most interesting that a symmetrical increase in the surface free energy of the i-PP gel crystals with increasing PE concentration (possibly due to entanglements of the PE chains with the i-PP crystals) is also obtained.

In conclusion, the deviation from the microhardness additivity law (Fig. 5.4, line 1) can be explained in terms of two distinct contributions: (a) a crystallinity depression caused by the coexistence of the PE and i-PP phases, which yields curve 2, and (b) a substantial decrease of the crystal microhardness of the PE and i-PP components caused by an increase of the surface free energy σ_e with composition, which leads to curve 3. It is suggested that the rise in σ_e rise is a consequence of the increase in the level of surface defects including entanglements (see Section 4.3).

5.1.3   Blends of non-crystallizable components

In addition to the studies on blends of polyolefins described above, in which the two components are crystallizable, investigations have also been carried out on blends of non-crystallizable components.

Amorphous films of PVF$_2$/PMMA blends were prepared by initial precipitation from a solvent and rapid solidification at $\sim$15 °C from the molten state (Martínez-Salazar et al., 1991). Moreover, these two constituents were considered as miscible (Noland et al., 1971). The PVF$_2$/PMMA compositions studied were 25/75, 45/55, 50/50, 55/45, 60/40 and 75/25 by weight. The presence of a single X-ray halo as well as a single T_g value for all the blends, in the above compositions range, favours the view that these materials are composed of homogeneous mixtures at molecular level. The parallel decrease of the microhardness, which obeys a simple expression:

$$H_{blend} = H_{PMMA}(1 - w_{PVF_2}) \tag{5.7}$$

w_{PVF_2} being the PVF$_2$ concentration) and the glass transition temperature, T_g, following the predictions of Gordon & Taylor (1952), suggests that the depression of microhardness is caused by the shift of T_g towards lower temperatures. It has been pointed out that the effect of the PVF$_2$ molecules is to act as a softening agent within the PMMA component.

5.2 Microhardness of coreactive blends: the influence of chemical interactions

Blends of condensation polymers are useful in a broad range of applications. An inherent property of such systems is the occurrence of transreactions – transesterification, transamidation, ester–amide interaction, depending on the chemical structure of the homopolymers blended. It is widely recognized (Kotliar, 1981; Devaux *et al.*, 1982; Fernández-Berridi *et al.*, 1995; Fakirov, 1999) that the said reactions (usually performed in the melt) lead, as a rule, to the formation of copolymers. Variation of the reaction parameters, most often the temperature and reaction time, enables different products to be obtained ranging from diblock copolymers to more or less random ones. DSC traces show that with the progress of transreactions in binary blends comprising at least one semicrystalline homopolymer the crystallizability of the resulting copolymers changes drastically – from a melting point depression to a complete loss of the crystalline features of the system with the appearance of a single glass transition T_g between those of the starting homopolymers. Such effects were demonstrated by Kimura *et al.* (1984) for PET/polyarylate (PAr) blends, and by Fakirov & Denchev (1999) for other blends of condensation polymers.

On the other hand, it was pointed out by Lenz & Go (1973, 1974) that if some random polyester copolymers are heated below the melting point of the crystallizable component for finite times, changes of the thermal properties of the system take place, suggesting restoration of the crystalline structure. In subsequent studies of this issue the formation of non-equilibrium blocky molecules via transreactions was proposed to be the source of the phenomenon, being promoted by an active ester interchange catalyst. In a systematic study Fakirov *et al.* (1996a–b) followed the influence of transreactions upon the copolymer crystallizability in immiscible, miscible binary and ternary polyester/polyamide blends. DSC, WAXS and dynamic mechanical thermal analysis (DMTA) were primarily employed to study the interchain reactions. It was thereby demonstrated that in thermal treatment in the melt a blend of two or more condensation homopolymers transforms first into a block copolymer:

$$(A)_n + (B)_m \longrightarrow \cdots (A)_x - (B)_y - (A)_z \cdots \tag{5.8}$$

and that this, in turn, changes into a more or less random one as a result of interchain reactions:

$$\cdots AABBBBBBBBBBBAAAAAAB \cdots \longrightarrow \cdots ABABBABABAA \cdots$$

This effect is called melt-induced sequential reordering. The driving force for this transformation is supposed to be the resultant entropy increase. More about the chemical interactions in blends of condensation homopolymers and copolymers can be found in Fakirov (1999), a book devoted to these problems.

5.2.1   Coreactive blends of PET and PC

Polymer blends based on a polyester and a polycarbonate have been shown to be immiscible provided no transesterification reaction occurs (Porter & Wang, 1992). Heat treatment of the same blends yielded different degrees of compatibility depending on the temperature and duration of the treatment, as well as on the presence and type of catalyst. This method has been successfully used to increase the compatibility of different polymers with poly(bisphenol-A-carbonate) (PC).

The microhardness of coreactive blends of PET and PC has been investigated over the whole range of compositions (Baltá Calleja *et al.*, 1997a). The blends with the PET content varying between 0 and 100 wt% have been prepared by means of coreactive mixing. For this purpose, PET and PC pellets were mixed in the reactor in the temperature range 272–290 °C and a vacuum of 0.1–0.2 mm Hg pressure in the presence of magnesium hydrohexabutoxy-o-titanate as a catalyst (1.5 cm^3 in 400 g of the mixture). After 45 min, the blend was extruded and granulated.

For further thermal and mechanical characterization of the blends, films were prepared from the pellets via injection moulding followed by compression moulding. The compression moulding was performed at a pressure of 5 MPa at 220 °C for 10 s. After moulding, the films were cooled between two metal blocks for 2 min. Films of different starting PET/PC ratio were prepared and tested by WAXS and found to be completely amorphous.

From the experimental results (Fakirov, 1999) and taking into account the experimental conditions used for the coreactive blending (275–290 °C, a treatment time of 45 min, the presence of a transesterification catalyst), one has to assume the occurrence of intensive chemical interactions. It is also to be expected that these reactions will lead not only to the formation of copolymers, but will also result in a more or less complete randomization of the sequential order of the repeating units (eq. (5.8)). If this is the case, the initially two-component blends should be converted into one-component ones. The latter represent one-phase systems (copolymeric material) whenever randomization (i.e. amorphization) takes place. That this occurs can be concluded from the DSC curves shown in Fig. 5.6. In contrast to the homopolymer PET the blends do not crystallize and they exhibit only one glass transition temperature.

The conclusion that the starting PET/PC blends are converted via coreactive blending into a one-component material with only one phase (amorphous) has important consequences when the 'blend' is characterized with respect to its mechanical properties. Unfortunately, this peculiarity of such blends is frequently disregarded or underestimated, and this results in erroneous conclusions concerning, for example, the miscibility of the starting materials (Porter & Wang, 1992).

A second point concerns the manner in which the blend composition is expressed. When dealing with blends of condensation polymers that have been reacted and converted into copolymers, the copolymers being uniform with respect to the number of components (as well as with respect to the number of phases provided no phase separation via crystallization or dephasing has occurred), it seems reasonable to express the ratio of the components in mol% rather than in wt%. This reflects more realistically the composition of the system and, at the same time, using a mole ratio gives some idea of the character of the sequential order in the chains, assuming complete randomization has taken place. The fact that the mole ratio reflects the block length when complete randomization is achieved allows one to make direct conclusions about the crystallization capability of the copolymers obtained. For example, in the present case only the blend richest in PET (90/10) is potentially crystallizable. For the rest of the blends, the PET 'blocks' are too short to form

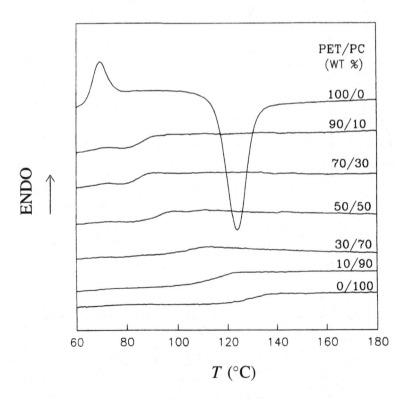

Figure 5.6. DSC curves for homo-PET and homo-PC and their coreactive blends of various compositions. (From Baltá Calleja *et al.*, 1997a.)

lamellae: the minimum thickness for lamellae is 50–60 Å. It should be emphasized that, in the case studied, the differences between wt% and mol% are not that big because the molecular weights of the two repeating units are rather close to each other, but there are cases in which the difference is significant.

Bearing in mind the outlined peculiarities of condensation polymer blends, and particularly when they consist of one component and one phase (this case is more the exception rather than the general rule since block copolymers usually consist of two, three, or more phases), the application of the additivity law for the evaluation of their characteristics does not seem to be completely justified. The observed good agreement between the measured microhardness values and the calculated ones (Fig. 5.7) allows one to make an important conclusion in this respect.

The solid straight line in Fig. 5.7 reflects the H values calculated according to the mechanical parallel model of eq. (5.1):

$$H = wH_a^{PET} + (1 - w)H^{PC} \tag{5.9}$$

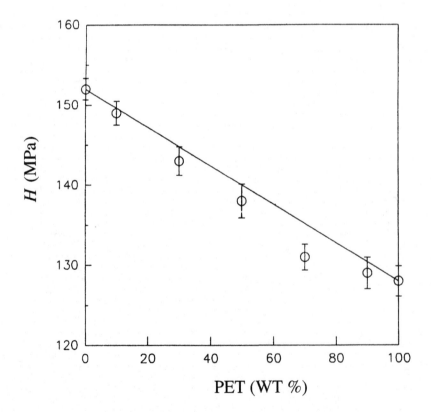

Figure 5.7. Dependence of microhardness H on compositions of coreactive PET/PCc blends. The solid line represents the H values calculated according to eq. (5.12). (From Baltá Calleja *et al.*, 1997a.)

where w and $1-w$ are, respectively, the weight fractions of PET and PC; and H_a^{PET} and H^{PC} are the microhardness values for amorphous PET and for PC, respectively.

Basically, the application of the additivity law for blends assumes the presence of spatially well defined regions of chemically and structurally uniform moieties, which is not the case for amorphous PET/PC copolymers since only a single T_g is observed. Then, it follows that the contribution of the two species to the microhardness of the copolymers is transferred via the repeating units constituting the copolymer molecules. Since in the present system there are no crystalline phases that can independently contribute to the microhardness of the copolymer, the observed microhardness can be regarded as arising from and depending on only the chemical composition of the copolymers. However, the microhardness can be considered as having a contribution from each of the copolymers. One can conclude that the additivity law can be applied at the molecular level, i.e. the microhardnesses of amorphous copolymers of differing composition obey an additivity law based on the microhardnesses of the respective amorphous homopolymers provided no other factors affect the microhardnesses of the copolymers.

It is important to note here that the additivity law is applicable to blends of miscible pairs of polymers. This was demonstrated for the blends of PMMA and PVF$_2$ discussed in the preceding section.

In conclusion, when working with blends of condensation polymers, one always has to take into account the possibility of chemical interaction and the formation of copolymers. The extent of this reaction is important because it is possible to obtain a one-component, as well as a one-phase, system when the blocky sequential order is converted to a random one. Such systems are very appropriate for the verification of relationships reflecting the effect of composition on various properties since they are free from other factors. Finally, in such cases one is dealing with copolymers distinguished by the creation of new chemical bonds, not with blends, although initially two or more homopolycondensates are mixed.

5.2.2 Blends of PET and PEN

The effect of chemical interactions on microhardness has been examined in another interesting system – blends of PET with PEN. Baltá Calleja *et al.* (1997b) indicated that in blends of PET and PEN their thermal behaviour strongly depends on the time during compression moulding (t_m) before the quenching of the films in ice-water. For t_m values between 0.2 and 0.5 min two glass transitions were observed by means of dynamic mechanical analysis, indicating the presence of two phases: an amorphous PEN-rich phase and an amorphous PET-rich phase. For t_m in the range 2–45 min only a single T_g and, consequently, a single phase was found to exist. Within this time range transesterification of the two compounds also takes place resulting in a copolyester of PET and PEN. The variation of the mechanical properties (microhardness) of these PET/PEN blends was therefore examined as a function of both composition and melt pressing time t_m.

Blends of these starting materials were obtained by coprecipitation from solution in hexafluoroisopropanol. Amorphous films were then obtained from the precipited powder by melt pressing *in vacuo* for different times t_m, varying from 0.2 min to 45 min followed by quenching in ice-water. In this way PET/PEN blends with weight compositions 90/10, 70/30, 60/40, 44/56, 30/70 and 10/90 were prepared.

The PET and PEN blends are completely amorphous after quenching from the melt as revealed by the DSC experiments (Zachmann *et al.*, 1994). Figure 5.8 shows in detail the variation of the microhardness with melt-pressing time t_m for the PET/PEN composition 44/56. For all compositions, H shows first a rapid initial increase with t_m, exhibiting a maximum just before $t_m = 10$ min and, then, for longer times, a gradual decrease down to values which can be even lower than the starting ones.

In order to explain the variation of H *vs* t_m it is convenient to analyse microhardness as a function of composition. It is found that for $t_m \sim 0.2$ min, H increases

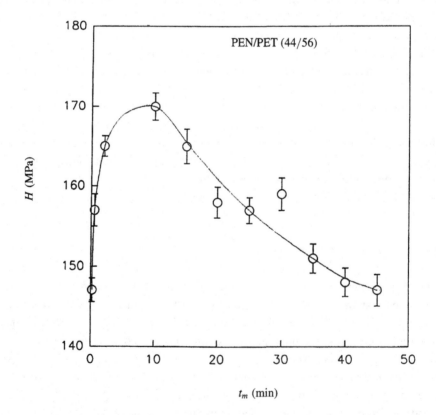

Figure 5.8. Microhardness H variation of a PET/PEN (44/56 by mol) blend *vs* melt-pressing time t_m. (From Baltá Calleja *et al.*, 1997b.)

linearly with increasing concentration of PEN content according to the prediction of the mechanical parallel model given by

$$H = H_a^{PET} \, w_{PET} + H_a^{PEN} \, w_{PEN} \qquad (5.10)$$

where H_a^{PET} and H_a^{PEN} are the microhardnesses of amorphous PET and PEN, respectively, and w_{PET} and w_{PEN} are the weight fractions of PET and PEN, respectively. These results indicate that the microhardness of the PET/PEN blends shows similar values to those obtained for PET/PEN amorphous random copolyesters as demonstrated by Santa Cruz *et al.* (1992). With increasing t_m up to 10 min one observes a shift of the straight line defined by eq. (5.10) towards higher microhardness values. Finally for $t_m = 45$ min one observes the lowest values of H.

Let us recall that when the melt-pressing time is about 0.2–0.5 min two T_gs are observed, indicating that there are two phases present. In case of $t_m \geq 2$ min a single T_g and, thus, a single phase is found. For $t_m \sim 10$–45 min no crystallization and melting during heating in the differential scanning calorimeter at $10\,^{\circ}\mathrm{C}\ \mathrm{min}^{-1}$ are observed, indicating that an amorphous copolyester has been obtained by transesterification of PET and PEN during melt pressing. The initial increase in microhardness up to $t_m = 0.5$ min could be due to the corresponding shift of T_g toward higher temperatures. It is known that in the case of amorphous blends temperature is the dominant parameter in determining the yield behaviour of the glassy material. The further increase in H when t_m reaches 2 min could be associated with the change from a two-phase system to a single amorphous phase composed of interpenetrating molecules of both polymers. Such an homogeneous system should offer a higher mechanical resistance to yield and to plastic deformation. Finally, H further increases with t_m up to $t_m \sim 10$ min at which point the copolyesters of PET and PEN have been formed by transesterification.

One may ask at this stage, why does H then gradually decrease with increasing t_m if the molecular weight and the viscosity remain practically constant as shown by additional measurements? One possible explanation is that at the beginning of the transesterification process the copolyester has a rather block-like character. Only after longer times does it become a statistical copolymer as found for other similar blends (Fakirov & Denchev, 1999). The results, therefore, indicate that the microhardness of the block copolyester is larger than that of the statistical copolymer. The existence of blocks may lead to a microphase separation between PEN and PET blocks. It seems, then, reasonable to assume that parallel packed sequences of blocks with the same chemical compositions would yield mechanically less easily than parallel copolymer sequences of statistical composition.

In conclusion, in order to obtain the optimum mechanical properties of these blends, one should use melt-pressing times in the range 5–19 min. Otherwise, the mechanical properties represented by microhardness may be reduced by 10–15%.

5.2.3   Blends with functionalized polyolefins

Chemical interactions are used for alloying even in blends of polyolefins provided the polyolefin is additionally functionalized. Thus, the preparation of polymer alloys by reactive blending has been shown to be a powerful method which brings together the advantageous properties of the starting components and minimizes their disadvantages (Lambla *et al.*, 1989). Formulation rules and the process technology for developing injection-moulded *i*-PP/polyamide (PA66) alloys and thermoplastic elastomers as well as the property profiles of the resulting products have been reported by Fritz *et al.* (1995). If PA66 is present in the blend as a finely dispersed phase, and is covalently linked to chemically modified *i*-PP via linkage molecules, then a stable two-phase material results. Such blends may be created from a binary system in which the entire *i*-PP matrix is subjected to a grafting reaction prior to the alloying step (Park *et al.*, 1990). Maleic anhydride (MAH) which is a highly efficient and well known linkage molecule has been used to initiate radically the functionalization of *i*-PP. These types of polymer blend absorb relatively little water, show an increased stiffness as compared to conventional blends and have a low cost. Their mechanical properties and limiting temperatures against heat distortion depend to a high degree on the resulting blend morphology. *i*-PP/PA66 compositions between 100/0 and 50/50 by weight using functionalized *i*-PP with various degrees of mainchain grafting, have been investigated using X-rays and microhardness. It was found that the microhardnesses of the injection-moulded reactive *i*-PP/PA66 blends can be described in terms of at least four contributions: (*a*) the crystalline α phase which shows a *c* axis molecular orientation parallel to the injection-moulding direction; (*b*) the amorphous *i*-PP phase which is preferentially located within the stacks of lamellae which are perpendicular to the injection direction; (*c*) the crystalline PA66 phase which is unoriented and does not show any apparent regular lamellar periodicity; and (*d*) the amorphous glassy PA66 phase. It has been shown that the presence of PA66 throughout the range of blends inhibits the crystallization of *i*-PP and induces a depression of crystallinity in the form of amorphous regions created outside the *i*-PP lamellar stacks, causing a depression of the microhardness compared to that derived from the additivity of the microhardnesses of the single components. Factors such as the main-chain grafting ratio and the degree of linkage between *i*-PP and PA66 chains can strongly affect the nature of the amorphous network and influence the mechanical yielding of the material as revealed by micro-hardness (Fritz *et al.*, 1995).

5.3   **Mechanical studies of condensation copolymers**

An interesting class of condensation copolymers is the liquid crystalline polymers (LCP). From the viewpoint of practical applications, LCP are very attractive as potential high-performance materials for use in high-modulus engineering plastics and fibres or for the production of displays and other optoelectronic devices. In recent years interest in rigid chain thermotropic LCP has grown considerably. Of particular interest has been the copolyester prepared from *p*-hydroxybenzoic acid (PHB) and 2-hydroxy-6-naphthoic acid (PEN) which form thermotropic liquid crystalline melts which can readily be processed into highly oriented fibres or films.

Random copolymers of PET and PHB and of PEN and PHB form thermotropic liquid crystalline melts if their PHB content exceeds 30% (Buchner *et al.*, 1988). In PEN/PHB copolyesters containing up to 50% PHB, crystals of PEN are formed, whereas in systems containing 80–90% PHB, crystals of PHB are found. The mechanical properties of these materials have been investigated as a function of composition using the microhardness technique (Baltá Calleja *et al.*, 1991). The mechanical behaviour has been interpreted in the light of several microstructural parameters, including the thickness of the crystals, polymorphic crystalline forms and the fraction of crystalline content. In the case of quenched glassy copolyesters containing PHB, the additivity of microhardness for the two single components was postulated. PHB and copolymers containing more than 50% PHB could not, however, be solidified in the fully amorphous form. For the crystallized samples of PET/PHB and PEN/PHB it was shown that the microhardnesses of the PET and PEN crystals were an increasing function of the crystal thickness. On the other hand, for high concentrations of rigid PHB units, the samples always crystallized and the material showed a microhardness increase which was proportional to the PHB content.

In summary, the microhardness of PET/PHB and PEN/PHB quenched copolymers show minima for blends of 30/70 and 50/50, respectively, due to the fact that the samples are fully amorphous for these compositions. If the PET, or the PEN, content is increased, the microhardness will increase owing either to the additive microhardness behaviour of the single components for amorphous samples, or to the increasing crystal thickness in the case of crystallized samples. If, on the other hand, the PHB concentration is increased the samples always crystallize and the crystallite-reinforced material shows a microhardness increase which is proportional to the PHB content of the crystal (Baltá Calleja, 1991).

The microhardnesses of a series of random copolyesters of 4-hydroxybenzoic (HBA) and 2-hydroxy-6-naphtoic acid (HNA) have also been investigated as a function of composition and temperature by Flores *et al.* (1997). The results reveal that, at room temperature, the microhardness of non-oriented materials deviates from the linear additivity of the microhardness of single homopolymers. Such a deviation is shown to be mainly related to changes in the molecular packing of the rigid chains. This packing, and as a consequence microhardness, can be characterized by an average cross-sectional area which includes crystalline and non-crystalline regions.

Mechanical studies have also been performed on another non-LCP system – copolyesters of PET and PEN, where both copolymer units having flexible chain segments and lack a liquid-crystalline behaviour (Santa Cruz *et al.*, 1992). The whole range of PET/PEN copolymers can therefore be prepared in the amorphous state.

Samples of PET/PEN copolymers with 10, 20, 30, 50, 80 and 100 mol% PEN have been synthesized. Amorphous films of the samples were obtained by melt pressing above the melting point and quenching in ice-water. The samples were then crystallized by annealing the glassy materials at various temperatures. The degree of crystallinity was calculated from the amorphous density measured on quenched samples and from the crystal density derived from the crystal unit cell.

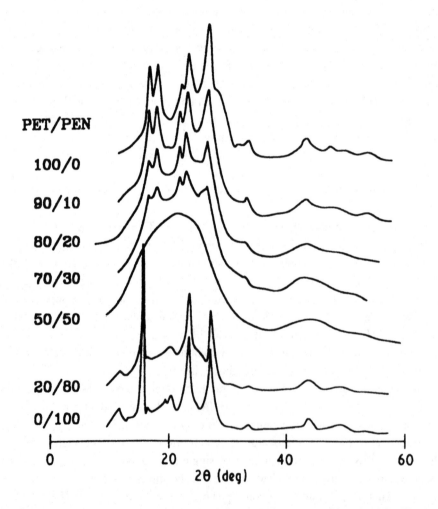

Figure 5.9. WAXS patterns of PET/PEN blends with a various compositions (mol%). (From Santa Cruz *et al.*, 1992.)

The whole range of PET/PEN copolyesters was completely amorphous when quenched from the melt as revealed by the WAXS patterns. However, when annealed at high temperature, some of the samples were capable of crystallizing. Figure 5.9 illustrates the WAXS patterns of the annealed copolyester series as a function of composition. For samples containing 0–30 mol% PEN, the PET sequences crystallize while the PEN segments remain in the amorphous regions. However, in the samples containing 80 and 100 mol% PEN, the PEN sequences crystallize in the α-polymorphic form. The PET segments are here excluded into the non-crystalline regions. Only samples containing 50 and 60% PEN (the latter is not shown in Fig. 5.9) and annealed at high temperature do not crystallize.

Figure 5.10 shows the typical variation of the microhardness of the quenched amorphous samples, H_a, and that of the crystallized samples, H_c, as a function of PEN content. It is seen that H_a increases linearly with the increasing concentration of PEN units, w_{PEN}, according to the predictions of the additivity model:

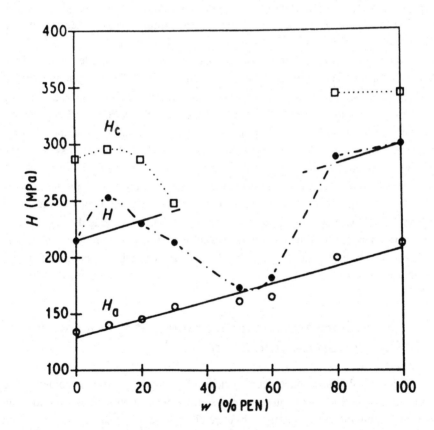

Figure 5.10. Experimental microhardness H data as a function of PEN content w for the quenched materials, H_a and the annealed samples, H. The crystal microhardness H_c was derived using eq. (5.15). (From Santa Cruz et al., 1992.)

$$H_a = H_a^{PET} \, w_{PET} + H_a^{PEN} \, w_{PEN} \qquad (5.11)$$

where w_{PEN} is the total concentration of PEN units within the copolymer. The larger values of H shown in Fig. 5.10 for the crystallized samples are related to the presence of either PET (left) or PEN (right) crystals. When the concentration of PET units is greater, the probability for bundles of PET sequences to agglomerate forming crystallites also increases. On the other hand, if the concentration of PEN segments is larger, then, the probability for these sequences to pack in the form of crystalline aggregates is also larger. Larger crystallinity and crystal thickness values consequently give rise to larger H values.

In addition, the microhardness of the crystals, H_c, was calculated using the additivity of crystalline and amorphous microhardness values (eq. (4.3)) for PET/PEN compositions of 100/0, 90/10, 20/80 and 0/100

$$H = H_c w_{c_L} + H_a (1 - w_{c_L}) \qquad (5.12)$$

where $w_{c_L} = \ell_c / L$ represents the so called linear crystallinity (weight fraction). Equation (5.12) has been shown to apply for PET samples crystallized at different temperatures and times of crystallization (Santa Cruz et al., 1991).

In conclusion, while in the case of PHB copolymers, samples with a high concentration of PHB units could not be quenched into the amorphous state, in the flexible copolyesters, glassy amorphous materials can be produced over the whole range of compositions. In the case of the amorphous PET/PEN copolymers the microhardnesses of the whole range of compositions follows a simple additive behaviour. Similarly to the results obtained in PET/PHB and PEN/PHB systems, the annealed samples show the lowest microhardnesses for concentrations near 50% (by mol). This finding is consistent with the fact that, even after annealing, the samples are always amorphous at these compositions. If, however, one increases either the PEN or the PET content, microhardness will always increase due to the contribution to the microhardness of the developing crystalline regions. The influence of crystal thickness, crystallinity and the microhardness of the crystals can be quantitatively accounted for.

5.4   Microhardness of multicomponent systems with T_g below room temperature

Micromechanical studies have been carried out on thermoplastic elastomers. The latter are a special class of multiphase systems (block copolymers) exhibiting an unusual combination of properties: they are elastic and at the same time tough and they show low-temperature flexibility and strength at relatively high temperatures (frequently ca 150 °C). In addition, they are easily processable. For this reason, they are nowadays of great commercial importance as engineering materials. In natural rubber and synthetic elastomers, the polymer chains are chemically linked into a three-dimensional network and they are not easily processable. In thermoplastic elastomers, however, the chemical cross-links are replaced by thermally labile tie points held together by physical forces: glassy, crystalline, or even hydrogen-bonded segments or ionic associations. In every case, they confer to the material a processability not found in the usual types of elastomer (Sperling, 1986; Legge *et al.*, 1987).

For a copolymer to behave as a thermoplastic elastomer the chains must contain two types of chain sequences: (i) amorphous, rubbery blocks (with T_g below room temperature) referred to as 'soft' segments, and (ii) hard blocks, typically with a T_g above room temperature and/or extensive crystallinity. The soft segments impart an elastomeric character to the copolymer and are bracketed by the 'hard blocks'. The hard segments can form intermolecular associations, resulting in a solid phase within a desired temperature range, thus imparting dimensional stability to the material. At elevated temperatures, dissociation of the physical bonds occurs, giving rise to processability. In order to meet the continuously increasing demands for thermoplastic elastomers with new or modified properties, copolymers with increasingly complex chemical composition are being synthesized. Representative thermoplastic elastomers are for instance the poly(ester ether) bases on poly(butylene terephthalate) (PBT) and poly(tetramethylene glycol) (PO4), the PBT sequences forming the hard segments, and the PO4 sequences the soft ones. The latter can be chemically modified by introducing polycarbonate moities in the soft segments (Sperling, 1986; Legge *et al.*, 1987).

Here we consider a series of new poly(ester ether carbonate) (PEEC) multiblock terpolymers with varying amount of ether and carbonate soft-segment content. Dielectric relaxation experiments on the same PEECs revealed the existence of two relaxation processes (Roslaniec *et al.*, 1995). The dielectric loss values show the existence of a relaxation maximum appearing at about 0 °C for 10 kHz (β relaxation) accompanied by a lower temperature relaxation (γ relaxation) which appears at about -50 °C.

The γ process, which appears at low temperatures, is due to local mode motions of the polar groups (ether and carbonate from the soft segments and ester from the PBT segments) in the amorphous phase. For low PC concentrations the γ process is mainly governed by the ether groups of the PO4 segments. As the number of PC soft segments increases, a shift of the transition temperature at fixed frequencies is observed and higher activations energies are measured. The second transition (β relaxation) is associated with the glass transition of the terpolymer. The dependence of T_g on the PC concentration suggests the existence of good mixing between PC and PO4 soft segments in the amorphous phase (Roslaniec *et al.*, 1995).

The microhardness of films of thermoplastic elastomers based on PBT–cycloaliphatic carbonate (PBT-PCc) block copolymers has also been studied (Giri *et al.*, 1997). The microhardness of the amorphous films has been discussed in terms of a model given by the additivity values of the single components H_a^{PBT} and H_a^{PCc}. It has further been shown that the deviation of microhardness from the additivity law, for the semicrystalline copolymers, is mainly due to the depression of the crystal microhardness of the PBT crystals and partly due to a decrease in crystallinity of the PBT phase. This study has been extended with the aim of following the effect of chemical composition, i.e. the introduction of PCc moities within the soft segments in place of polyether moities while keeping the PBT hard segments content constant, on the number of phases, and examining their transitions and mechanical properties by means of a joint study of DMTA, WAXS, SAXS and microhardness measurements (Baltá Calleja *et al.*, 1998).

Terpolymers with a varying content of PO4 and PCc segments, each between 0 and 40 wt%, and a constant PBT segment content of 60 wt% have been examined in

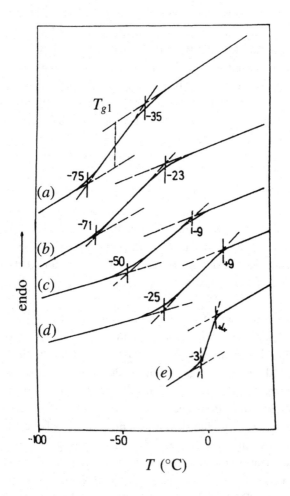

Figure 5.11. DSC curves in the low-temperature region for PBT/PO$_4$/PCc polyblock copolymers with various compositions (Baltá Calleja *et al.*, 1998) (see Table 5.3, for details of the compositions).

the form of thin films. Figure 5.11 shows DSC traces taken in the low-temperature range for polyblock copolymers of various compositions. The first and the last curves (curves (a) and (e)) correspond to the copolymers of the type $(AB)_n$ and are poly(ester ether) (PEE) and poly(ester carbonate) (PEC), respectively. The rest of the curves (Fig. 5.11 curves (b), (c) and (d)) refer to copolymers of the type $(ABC)_n$ and are poly(ester ether carbonate)s (PEECs). It is to be noted that with increasing PC content the glass transition interval becomes narrower and the glass transition temperature is shifted systematically to higher temperatures. The T_g values (evaluated at the temperature at which the change in heat capacity is $1/2$) together with other data are summarized in Table 5.3.

Taking into account the chemical composition of the copolymers under investigation as well as the fact that the T_g of the hard segments of PBT is around $50\,^{\circ}C$ one has to assign the observed glass transition temperatures (T_{g1}) to the amorphous phase arising from the soft segments (Table 5.3).

It is noteworthy that regardless of the chemical composition, i.e. whether one is dealing with a copolymer of the $(AB)_n$-type or of the $(ABC)_n$-type, one observes in this low-temperature region only one T_g indicating the presence of only one amorphous phase comprising the soft segments. This is the case for the samples PEE and PEC containing chemically uniform soft segments, PO4 or PCc (Fig. 5.11, curves (a) and (e), respectively). For the other cases, when the soft segments are no longer chemically uniform but are copolymeric, one has to assume that the presumably two amorphous phases (from PO4 and PCc) are miscible or that they do not phase separate during the treatment.

Table 5.4 shows the clear influence of PCc content on the microhardness. The dominating contribution of crystallinity in the mechanical properties of polymers is well known. Therefore, to understand better the correlation between the chemical composition and the microhardness of the copolymers one has first to examine the behaviour of the crystalline phase. Figure 5.12 shows the nearly parallel dependence of the microhardness and the crystallinity on the PCc content within the copolymers. H drops continuously with increasing PCc wt% almost in the same manner as the crystallinity does. Such a parallel behaviour of H and w_c shows that the effect of chemical composition on the microhardness is transferred via the crystallinity.

The plot of H vs w_c for copolymers with various compositions offers the opportunity to derive, by means of extrapolation, the values of H for a completely amorphous copolymer (H_a^{PEEC}) and for a copolymer with 60 wt% fully crystallized PBT component. The extrapolation of the straight line to $w_c = 0$ and 100% leads to the following values: $H_a^{PEEC} \simeq 6.0\,\mathrm{MPa}$ and $H_c^{PEEC} \simeq 106\,\mathrm{MPa}$. It is noteworthy that these values are significantly lower than those for the homopolymer PBT; the samples studied contained only 60 wt% of this polymer. Thus, the remaining molecular strains must belong to a much softer material which causes the drop in the H. Before discussing in more detail the microhardness behaviour of the system

Table 5.3. *Thermal properties of PEEC block terpolymers.*

Composition (wt%) (PBT/PO4/PCc)	Degree of polymerization of PBT	$[\eta]$ (dl/g)	First heating			Second heating				
			T_{g1} DSC	T_{g1} DMTA	T_{g2} DSC	T_c (°C)	ΔH_c (J g^{-1})	T'_m (°C)	T''_m (°C)	$\Delta H''_m$ (J g^{-1})
100/0/0			55							
60/40/0 (a)	7.7	1.16	−56	−50	57	171	39.7	170	201	39.7
60/32/8 (b)	8.4	1.19	−45	−34	55	145	25.1	146	179	31.8
60/20/20 (c)	9.4	1.16	−32	−17	54	111	20.2	108	151	19.0
60/12/28 (d)	10.1	1.04	−10	−3	53	70	20.2		131	18.1
60/0/40 (e)	11.1	1.02	1	19	52	56	13.9		102	12.8

Table 5.4. *Crystallinity w_c, crystal size values D_{hkl}, long spacing L, lamellae thicknesses ℓ_c and ℓ_w and microhardness H of PEEC block terpolymers.*

Composition (wt%) (PBT/PO4/PCc)	w_c (%)	WAXS D_{100} (Å)	D_{010} (Å)	L (Å)	SAXS ℓ_c (Å)	ℓ_w (Å)	H_{exp}	Microhardness (MPa) H_{cal}	$\Delta H = H_{cal} - H_{exp}$
100/0/0	59.0								
60/40/0	37.7	84.7	116.6	146.0	55.0	83.7	29.6	85.1	55.5
60/32/8	25.8	87.8	93.8	130.8	33.8	77.2	22.8	68.5	45.7
60/20/20	24.3	105.0	113.9	118.5	28.8	89.6	18.6	66.4	47.8
60/12/28	21.2	86.2	119.6	118.5	25.1	85.3	17.5	62.0	44.5
60/0/40	13.0	77.9	104.0	116.3	15.1	77.3	15.5	50.6	35.1

let us summarize the effect of chemical composition on the number of phases and transition temperatures.

The data obtained from calorimetric, dynamic mechanical and X-ray analysis (Table 5.3) indicate the existence in the copolymers of three phases – two amorphous and one crystalline. The hard PBT segments give rise to the crystalline phase and to one of the amorphous phases; the soft segments contribute to the second amorphous phase. These three phases are characterized by their transition temperatures T_m, T_{g2} and T_{g1}, respectively (Table 5.3).

The drastic increase in T_{g1} (by 50 °C) and the narrowing of the glass transition interval (from 40 to 7 °C) with increasing PCc content (Fig. 5.11 and Table 5.3) are a result of a decrease in the chain flexibility of the soft segments. However, the flexibility of the PBT chains remains unaffected by the presence of PCc moities, as can be concluded from the constancy of the T_g of PBT (Table 5.3, T_{g2}). In fact, what it is changed is the mobility (not the flexibility) in the PBT hard segments dispersed in the amorphous soft-segment matrix. The crystallization of PBT segments presumes such a mobility. In addition one has to take into account the fact that the hard segments are not only 'embedded' in the soft-segments matrix but they are also chemically bonded to each other by means of these soft segments with varying flexibility. The model of Fig. 5.13 attempts to illustrate the structure of hard PBT particles embedded in the amorphous phase of soft segments. It also shows the phase segregation in the amorphous regions where interconnection between soft and hard segments takes place.

By introducing PCc moities into the soft segments one decreases dramatically their flexibility thus restricting the mobility of the hard segments which, in turn, hampers the crystallization process. As a result, when the PCc content in the copolymers increases one can expect a decreasing degree of crystallinity and more imperfect crystallites as we actually observe (Table 5.4).

These considerations allow one to conclude that the effect of the PCc content on the crystallizability of the PBT hard segments is transferred by the drastic changes in the flexibility of soft segments interconnecting the crystallizable PBT hard segments. The strong influence of the PCc component on the flexibility of the soft segments is enhanced by another peculiarity of this amorphous phase. As we shall see below, the introduction into the PO4 soft segments of the chemically different component (PCc) does not yield a new amorphous phase, even if the PCc component dominates. One is dealing with a uniform amorphous phase regardless of the chemical composition of this phase. Nevertheless, its glass transition temperature, increases with the PCc content (Table 5.3, T_{g1}). This means that the PCc moities increase the rigidity of the soft segments by introducing chemical bonds with a higher rotation potential.

The conclusion drawn with respect to the effect of PCc content on the crystallinity is important for the interpretation of the mechanical properties.

Let us now return to the microhardness issue. The measured H values for the terpolymers studied (Table 5.3) are very low in comparison to the values known for common synthetic polymers even those in the amorphous state (Baltá Calleja, 1994). What is more, the H values are more than three times smaller than the ones for the same terpolymers calculated by means of the additivity law:

$$H = wH^{PBT} + (1 - w)H^{soft} \qquad (5.13)$$

where w is the weight fraction of the hard PBT segments.

Taking into account that the T_g of the soft-segment amorphous phase lies between -50 and $0\,°\mathrm{C}$, depending on the PCc content (Table 5.3), one can expect that at room temperature such a phase will be a very soft material much closer to a liquid than to a solid. For this reason one may expect that $H^{soft} \simeq 0$ and, as a result, that H will be depressed with increasing values of w according to the simple expression:

$$H = wH^{PBT} = w\left[H_c^{PBT} w_c + H_a^{PBT}(1 - w_c) \right] \qquad (5.14)$$

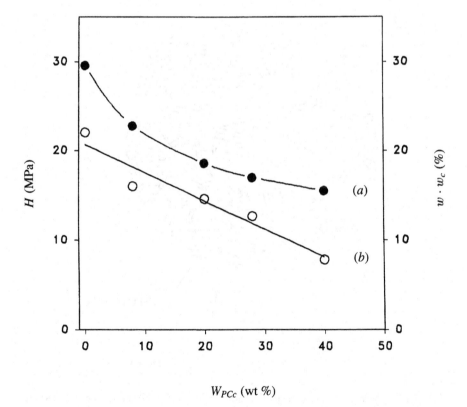

Figure 5.12. Variation of (a) microhardness and (b) total degree of crystallinity $w \cdot w_c$ as a function of PCc content, W in PEEC. (After Baltá Calleja et al., 1998.)

By applying the numerical values $w = 0.6$, $H_c^{PBT} = 287$ MPa and $H_a^{PBT} = 54$ MPa (Giri $et\ al.$, 1997) one can derive calculated values H_{cal} for the terpolymers depending on w_c (Table 5.4).

In order to explain such a microhardness depression, $\Delta H = H_{cal} - H_{exp}$ (Table 5.4), it is convenient to consider the behaviour of the harder component dispersed in the liquid component (with 'zero' hardness). It seems reasonable here to assume that the total microhardness of such a system will also depend on the viscosity of the soft component in which the particles of the harder component are 'floating'. The deviation of H_{exp} from H_{cal} may actually reflect the viscosity of the soft-segment phase which also contributes to the resistance of the total system against the applied load. The viscosity of the soft phase introduces two important effects.

The first one is related to the above mentioned 'floating' effect of the hard segments within the liquid matrix. Actually, the soft phase plays a plasticizing role strongly reducing H^{hard} depending on its viscosity. The lower the viscosity,

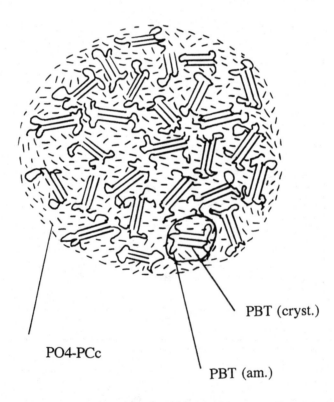

PO4-PCc

PBT (cryst.)

PBT (am.)

Figure 5.13. Schematic model showing the hard PBT particles (crystalline cores and surface amorphous regions) embedded in the amorphous phase of the soft segments. (From Baltá Calleja $et\ al.$, 1998.)

i.e. the lower the T_g^{soft} (T_{g1}), the stronger the plasticizing effect, i.e. the larger the depression in H^{hard}.

The second effect, caused by the viscosity of the soft phase, concerns the ability of the hard segments to crystallize. As discussed in the foregoing paragraphs the crystallization process presumes a mobility of the crystallizing segments. In the present case the hard segments are linked to each other through the soft segments. It is quite obvious that the higher flexibility of the soft segments (lower T_g^{soft}), and consequently the lower viscosity, will enhance the crystallization capability of the hard segments. At the same time the higher crystallinity results in higher H. It turns out that the viscosity of the soft amorphous phase creates two competing effects: on the one hand the plasticizing effect leading to a decrease in the overall microhardness H and, on the other, the enhanced crystallizability results in an increase in the overall microhardness H.

5.4.1   Systems which deviate from the hardness additivity law

In the preceding section it was demonstrated that multiphase systems comprising a phase with T_g below room temperature are distinguished by peculiarities in their deformation mechanism. While in the same subsection it was attempted to account for this peculiarity qualitatively, in this section it will be demonstrated that by means of eq. (3.4), which expresses the relationship between T_g and the microhardness, it is possible to describe the behaviour of the system in a quantitative manner.

According to the additivity law, eq. (1.5), one can calculate the microhardness H of any multicomponent and/or multiphase system provided the microhardness of each component and/or phase H_i and its mass fraction w_i are known. This relationship is of great value because it offers the opportunity to characterize micromechanically components of a system which are not accessible to direct measurement.

In the preceding chapters it has been demonstrated that many semicrystalline polymers, copolymers and blends obey the additivity law. Exceptions, such as blends of HDPE with PP are explained by a peculiarity in the morphological structure of the crystallites formed (mostly related to the surface energy) (see Sections 4.3 and 5.1.2).

The same approach applied to thermoplastic elastomers of the PEE-type fails to explain the large discrepancy (up to 100 MPa when the measured H values are in the range 20–40 MPa) between the experimental values and those calculated according to eq. (3.4) (Fakirov et al., 1998; Apostolov et al., 1998). For this reason one has to look for other factors which may be responsible for such a discrepancy. Before disclosing these let us recall some of the characteristic features in the structure and morphology of thermoplastic elastomers of PEE-type which are closely related with the problem discussed.

As pointed out before, themoplastic elastomers of PEE represent polyblock copolymers comprising PBT as the hard segment and poly(glycols) as soft segment.

The latter could be poly(tetramethylene glycol) (PTMG) or poly(ethylene glycol) (PEG). On the laboratory scale, as mentioned above, PEECs, which have also PCc as their soft segments, have been synthesized (Roslaniec et al., 1995).

Starting from a basic knowledge of PEE-type thermoplastic elastomers (Schroeder & Cela, 1988; Legge et al., 1987), as well as, the thorough structural characterization of the copolymers, with PTMG and PCc or only PEG (Fakirov et al., 1990; Stribeck et al., 1997) as soft segments, it is concluded that they are multiphase systems. Basically, they consist of at least four phases: crystalline PBT, amorphous PBT, crystalline PTMG (PEG) and amorphous PTMG (PEG). Since the polyglycol incorporated in the copolymer chains melts below room temperature one is dealing in practice with three phases. These phases are very well expressed on the DSC traces: a melting peak from the PBT crystals and two glass transition temperatures: a T_g around $-50\,^{\circ}$C for the soft segments of only PTMG or PEG (for PEEC, a T_g up to $1\,^{\circ}$C depending on the PCc amount, Table 5.3) and another T_g around $50\,^{\circ}$C arising from the amorphous PBT regions. (The structural model of these systems is depicted in Fig. 5.13.)

As emphasized above, in contrast to common thermoplastics, thermoplastic elastomers contain a very soft phase (with T_g around $-50\,^{\circ}$C), which is in a liquid state at room temperature and is characterized by a viscosity closer to that of low-molecular-weight liquids rather than a solid amorphous polymer. In this respect it seems useful to recall that the molecular weight of the PTMG and PEG used is 1000, i.e. one is dealing with typical oligomer systems. For this reason it looks reasonable to accept that such a liquid will be characterized by a negligibly small microhardness, H^s, in the equation:

$$H = w \left[H_c^h \, w_c + H_a^h (1 - w_c) \right] + (1 - w) H^s \tag{5.15}$$

where w is the mass fraction of hard segments (PBT in the present case), H_c^h and H_a^h are the microhardnesses of the crystalline and amorphous phases, respectively, and w_c is the degree of crystallinity of PBT.

Assuming $H^s = 0$ the calculations of H according to eq. (3.4) for a series of PEEs lead to a discrepancy between the measured H_{exp} and calculated H_{cal} amounting to 40–64 MPa, depending on the soft-segment composition as will be discussed below.

What could be the reason for the failure of the additivity law? Obviously one has to assume that, for multicomponent and/or multiphase systems, when one of the components (phases) is characterized by a viscosity at room temperature which is typical for low-molecular-weight liquids, the microhardness behaviour of the entire system should be different from the case in which all the components (phases) have T_gs higher than room temperature because the mechanism of the response to the applied external mechanical field is different. In the latter case all the components (phases) plastically deform as a result of the applied external force. In the former case, in addition to the plastic deformation of the harder components (phases) they are also displaced within the soft (liquid) matrix in which they are 'floating'. The extent of this displacement depends on the viscosity of the matrix (the softer component and/or phase). For this reason the harder components cannot display their inherent microhardness. The microhardness is reduced by the ability of the harder components to move.

The question arises: how one can account for this microhardness depression effect? As demonstrated above, the simple assumption that the soft segments have $H \simeq 0$ does not solve the problem. It is necessary to characterize the ability of the harder phase to move about within the soft matrix, and this will depend on the viscosity of the matrix, i.e. the soft-segment phase in the present case. Since T_g and viscosity are closely related to each other one can look for an analytical relationship between microhardness of amorphous polymers and their T_gs.

For a quantitative evaluation of the microhardness depression effect one has to replace H^s in eq. (5.15) with eq. (3.4) using for T_g the glass transition temperatures of the soft-segment phase T_g^s. This leads to the expression:

$$H = w\left[H_c^h w_c + H_a^h(1 - w_c)\right] + (1 - w)(kT_g^s + C) \tag{5.16}$$

Calculation of H for PEE and PEEC by means of eq. (5.16) offers data which are in a good agreement with the measured values H_{exp} as shown in Table 5.5 (samples 1–6).

Since the linear relationship between H and T_g (see Chapter 3) seems to be valid over a rather wide temperature range (in the present case it is proven for T_g values between -50 and $250\,°C$ at least), eq. (5.16) can be rewritten in such a way that it also accounts for the amorphous hard-segment phase:

$$H = w\left[H_c^h w_c + (1 - w_c)(kT_g^h + C)\right] + (1 - w)(kT_g^s + C) \tag{5.17}$$

where T_g^h is the glass transition temperature of the amorphous hard-segment phase. The H_{cal} values obtained are very close to those derived from eq. (5.16) and given in Table 5.5. The excellent agreement found between the H_{exp} and H_{cal} is not surprising when eq. (5.17) is applied, since the contribution of the amorphous hard-segment phase has already been accounted for in deriving the linear relationship between H and T_g (PBT was one of the polymers used for the plot in Fig. 3.11).

The advantage of the modified additivity law incorporating T_g (eq. (5.17)), is that it is possible to use it to account for the contribution of any amorphous phase and/or component to the overall microhardness of the system, provided the T_g of this phase and/or component is known. Hence, for systems which contain more than one crystalline and/or amorphous phases with glass transition temperatures and mass fractions T_{gi} and w_i, respectively, the additivity law can be presented in the following way:

$$H = \sum_i w_i H_{ci} w_{ci} + \sum_i w_i(1 - w_{ci})(1.97T_{gi} - 571) \tag{5.18}$$

Table 5.5. *Composition, pretreatment, crystallinity, glass transition temperature T_g^s and experimental and calculated microhardness values, H_{exp}, and H_{cal} and their difference $\Delta H = H_{cal} - H_{exp}$ for thermoplastic elastomers of PEE- or PEEC-type.*

Sample no	Copolymer	Composition (wt%)	Treatment Drawing λ	Treatment Anneal. (°C)	Crystallinity (%)	T_g^s (°C)	H_{exp}	Microhardness (MPa) H_{cal} (eq. (5.16))	ΔT	H_{cal} (eq. (5.17))	ΔT
1	PBT/PTMG	60/40		70	37.7	−56	29.6	85.7	56.1	29.7	0.1
2	PBT/PTMG/PCc	60/32/8		70	25.8	−45	22.8	69.5	46.7	21.8	1.0
3	PBT/PTMG/PCc	60/20/20		70	24.3	−32	18.6	59.9	41.3	22.0	3.4
4	PBT/PEG	57/43	5	170	41	−44.5	34.2	85.2	51.0	34.4	0.2
5	PBT/PEG	57/43		25	37	−41	30.7	79.9	49.2	31.6	0.9
6	PBT/PEG	57/43		170	41	−434	32.9	85.2	52.3	35.4	2.3
7	PBT/PEG	75/25		25	35	−16.9	47.3	101.7	54.4	85.1	37.8
8	PBT/PEG	75/25		150	39	−34.4	44.2	108.5	64.3	83.7	39.5

In this form, in contrast to the traditional one (eq. (1.5)), the additivity law is applicable to multicomponent or multiphase systems comprising liquid-like components or phases displaying a more complex deformation mechanism than the case in which all the amorphous components have a T_g above room temperature.

5.4.2 Some limits of the equation $H = kT_g + C$

Let us return to Table 5.5 in which data for other PEEs with not such a good agreement between H_{exp} and H_{cal} (according to eq. (5.16)) (samples 7 and 8) are presented. A possible explanation for the different behaviour of these two PEE samples is the difference in their composition. Samples 1–6 are characterized by hard-/soft-segment ratios of roughly 60/40 while in the samples 7 and 8 this ratio is 75/25. The fact that in the second case the hard PBT segments dominate posits another response mechanism to the mechanical field – the PBT hard segments are no longer 'floating' in the liquid-like matrix of soft segments.

5.5 Microhardness of polymer composites

Composites are defined as materials consisting of two or more distinct phases with a recognizable interface (Hull, 1981). This definition is generally restricted in practice to materials containing fibrous reinforcements or reinforcements with different length and cross-section dimensions (described by the aspect ratio, e.g. platelet or flake), which are embedded in a continuous matrix. Incorporating this reinforcement improves the mechanical performance. That is the basic difference between reinforced and filled systems in which fillers are used to reduce cost but often degrade instead of improve the mechanical property. The distinction between reinforced and filled systems is sometimes unclear (Karger-Kocsis, 1996).

Most naturally occurring materials derive their salient properties from a combination of two or more components which can be readily distinguished when examined in light or electron microscopes. Thus, for example, many tissues in the body that have high strength combined with enormous flexibility are made up of stiff fibres such as collagen embedded in a lower-stiffness matrix. The fibres are aligned in such a way as to provide maximum stiffness in the direction of high loads and are also able to slide past each other so that the tissue is very flexible. Similarly, a microscopic examination of wood and bamboo reveals a pronounced fibrillar structure which is very apparent in bamboo when it is fractured. It is not surprising that bamboo has been called 'nature's fibre glass' (Hull, 1981).

A peculiarity of polymeric composites that makes them very attractive is that their structure and thus their mechanical property profile can be tailored to given requirements and service conditions. Composite materials have been classified in many ways depending on the ideas and concepts that need to be identified. For the purposes of the present presentation only a classification emphasizing the size of the reinforcing elements will be used. With this in mind, three basic groups have to be mentioned: (i) continuous fibres with a thermosetting or thermoplastic matrix; (ii) short fibres with a thermoplastic matrix, and (iii) 'molecular' composites (LCP with a thermoplastic matrix).

5.5.1   Carbon fibre composites

In the addition of the classical reinforcement material – glass fibres – to polymers, the carbon fibre (CF) dramatically increases many of the polymers' mechanical properties (Fitzer, 1985; Savage, 1993). In many cases, these composites have become structural substitutes for metals, offering increased strength, lower thermal expansion and higher corrosion and fatigue resistance, all at lower weights. In aerospace applications, advanced composites containing up to 60 vol% CFs are increasingly used in structural applications. While many of the mechanical properties of CF reinforced composites (CFRC) have been well characterized, often by destructive testing, microhardness determination of the composites, a non-destructive technique, has not been commonly used. Paplham *et al.* (1995) investigated the microhardness of various composites containing different CFs and resins. The materials examined included an epoxy, similar to those used in aerospace structural applications, and an aromatic thermoplastic polyimide (TPI) that has promising mechanical and dielectric properties, high-temperature stability, solvent resistance and processability. These studies indicated that CFRC cover microhardness ranges between polymers and hard metals.

Both thermoset and thermoplastic resins and CF composites have been examined using microhardness techniques. The thermoset resin used was an epoxy, both with and without PA6 particles that served as a toughening agent.

Figure 5.14 shows the microhardnesses for the resins and for their composites. The microhardness perpendicular to the fibres is proportional to the yield stress required to permanently deform the material, while the microhardness parallel to the fibre is more affected by elastic modes of deformation. There is also a great variation in microhardness in the composites depending on the localized resin concentration. The limits of variation are shown in Fig. 5.14 with lower microhardness values indicating a resin-rich area, and higher microhardness values indicating that less resin is present. The limits shown are one standard deviation from the mean.

Firstly, we discuss the microhardness values obtained for the epoxy materials first. The addition of the CFs greatly increased the microhardness ($\simeq$740 MPa) with respect to that of the resin ($\simeq$314 MPa). This difference reflects the tremendous improvement in mechanical properties resulting from the presence of reinforcing CFs. In addition, the PA6 particles added for toughening induce a much higher increase in the microhardness in the composite.

This increase in the microhardness resulting from CF reinforcement has been also seen in thermoplastic TPI resin. In addition, microhardness is also influenced by the presence of crystallinity.

While the average values for the microhardness of all the TPI resin composites are similar, the large deviation in individual microhardness measurements, for both the TPI and epoxy systems, is of interest. The presence of resin-rich areas, resulting from inadequate surface resin removal or from uneven resin impregnation during manufacture, has a significant effect on the local microhardness. The shape of the measured indentation also varies greatly, depending on the fibre location with regard to the tip of the diamond. Impressions often end abruptly at the edge of a fibre, sometimes to reappear on the other side. This indicates elastic recovery mechanisms due to fibres that are not present in systems of just resin. In future it may be possible to use the shape of the indentation as a means of understanding the load transfer between the fibre and the matrix and the mechanisms for impact resistance in composites.

The high microhardness values found for these composites illustrate that the gap in microhardness between polymers and hard metals has been narrowed (see Fig. 1.2). While no microhardness values are available for CFs, the microhardness values for reinforced materials, especially high local values for areas with little resin, indicate the microhardness of CFs should be at least 1000 MPa. This agrees with the published microhardness for graphite (1 GPa) (Skinner & Gane, 1973), which

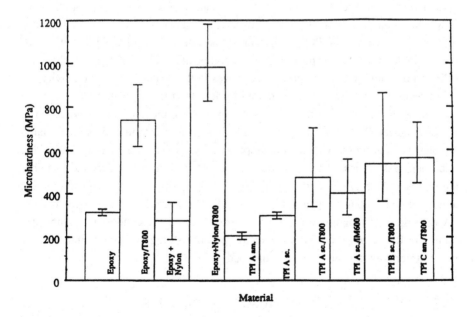

Figure 5.14. Microhardness for the amorphous (am) and semicrystalline (sc) resins and their composites. The bars denoted the standard deviation. (From Paplham *et al.*, 1995.)

should probably be the limiting lower value for the microhardness of CFs, and for diamond (7 GPa), which should represent the upper limit (Riedel, 1992).

In conclusion, the localized nature of the microhardness test gives information regarding heterogeneity that is often not available with other analytical techniques. However, care must be taken when utilizing microhardness to predict bulk mechanical behaviour, as a high number of measurements may be required to determine an average value that is a true representation for the material. These studies demonstrate the validity of this testing method in comparing different systems, from just resin to composite (Paplham et al., 1995).

5.5.2   Sintered materials

Another polymer composite that is of interest is the more complex system: SbCl$_5$–doped polyparaphenylene/polyparaphenylene sulphide (PPP/PPS) sintered materials (Rueda et al., 1988).

Microindentation hardness has been used to characterize the surface mechanical behaviour of two series of sintered PPP/PPS composites over a wide range of compositions.

Composites of PPP with PPS have been used in an attempt to improve the processability of conducting PPP. PPS is a versatile commercial material and its mechanical properties are partly retained when blended with PPP. The microhardness of PPP/PPS sintered composites can be described in terms of an additive system of two independent components H_1 and H_2 for weight concentrations of PPS lower than 70%. For PPS weight concentrations larger than 70% the presence of micropores within the PPS component causes deviations of the H additivity of the single components. Exposure of these composites to an SbCl$_5$ atmosphere for PPP compositions greater than 20% improves both the electrical conductivity, up to the level of 1 Ω^{-1} cm^{-1} (as reported by Rueda et al. (1987)), and the surface microhardness, to values of $\sim$175 MPa, approaching those of some metals. The use of annealed PPP in the sintered composites gives rise to materials with a large number of micropores which after doping show similar conductivity levels and microhardness values. These values are not restricted to the material surface but are representative of the material volume. Exposure of the composites to ambient atmosphere after long storage times (several days) reduces the conductivity level by several orders of magnitude. The microhardnesses for the conducting composites quoted in Rueda et al. (1998) were measured after storage for one month and are much larger than the values for the undoped materials.

5.5.3   Fullerene–polymer matrix composites

The microhardness of films of fullerene–PE composites prepared by gelation from semidilute solution, using ultra-high-molecular-weight PE (6×10^6), has been also

determined (Baltá Calleja *et al.*, 1996). These composite materials were characterized by light microscopy and X-ray diffraction techniques. The microhardness
of the films is shown to increase significantly with the concentration of fullerene
particles in the films. For instance, by adding about 2.5 wt% fullerene, one obtains
an increase in hardness of the original composite of about 30%, i.e. a value as high
as that of glassy PET. In addition, annealing at high temperature (130 °C) for long
times (10^5 s) leads to H values of $\sim$190 MPa, which are beyond the theoretical
limit for fully crystalline PE. The microhardening of the composites with annealing
temperature has been identified with a thickening of the PE crystalline lamellae.

Comparison of X-ray scattering data and the microhardness values after annealing leads to the conclusion that there is phase separation of fullerene molecules
from the PE crystals within the material. Fullerene–PE composites exhibit an
unexpectedly large microhardness increase as the temperature is increased above
75 °C and this has been ascribed to the hardening of fullerene aggregates within the
composite (Baltá Calleja *et al.*, 1996).

5.6 Mechanical model of microfibrillar-reinforced composites

A new type of composite material starting from polymer blends has been developed.
Due to the fact that the reinforcing elements are the basic morphological entities
of oriented polymers, the microfibrils, these new composites have been named
microfibrillar-reinforced composites (MFC) (Evstatiev & Fakirov, 1992). MFC,
however, clearly differ from traditional composite systems. Since the microfibrils
are not available as a separate component, the classical approach to composite
preparation is inappropriate for MFC manufacturing.

MFC are prepared from polymer blends of immiscible partners. There also needs
to be a significant difference in the melting points, T_m, of the constituent polymer
blends. The essential stages of MFC preparation are as follows: (i) blending, (ii)
extrusion, (iii) drawing (with good orientation of all components), and (iv) annealing
at constant strain above the T_m of the lower-melting component and below the T_m of
the higher-melting one. During the drawing step, the blend components are oriented
and microfibrils are created (the fibrillization step). In the subsequent annealing process, in which melting of the lower-melting component occurs (isotropization step),
it must be guaranteed that the oriented microfibrillar structure of the higher-melting
component is preserved (Fig. 5.15(*a*)). It is important to note here that MFC are
based on polymer blends but they should not be considered as 'drawn blends' since
the isotropization step results in the formation of an isotropic matrix reinforced by
microfibrils of the higher-melting component, i.e. one is dealing finally with a typical composite material. The mechanical parameters of MFC (Young's modulus and
tensile strength) are higher by 30–50% than the average values of the components

and comparable to those of short glass-fibre-reinforced polymer composites having the same matrix.

When MFC are prepared from blends of condensation polymers as a result of chemical reaction (additional condensation and transreactions) taking place at the

(a)

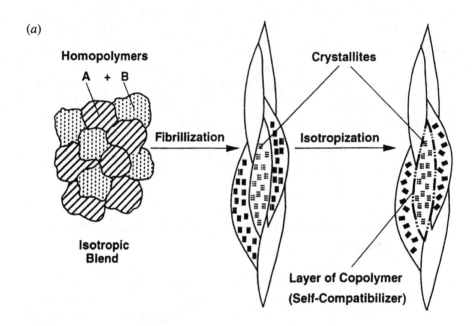

(b)

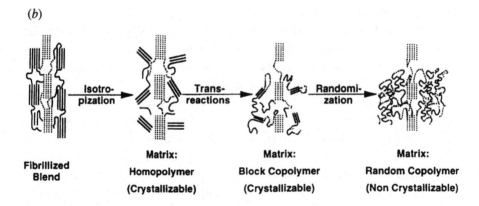

Figure 5.15. MFC can be obtained from incompatible polymer blends by extrusion and orientation (the fibrillization step) followed by thermal treatment at a temperature between the melting points of the two components at constant strain (the isotropization step). The block copolymers formed during the isotropization (in the case of condensation polymers) play the role of a self-compatibilizer. Prolonged annealing transforms the matrix into a block and thereafter into a random copolymer: (a) an MFC on the macro level, (b) an MFC on the micro (molecular) level (Fakirov & Evstatiev, 1994).

fibril/matrix interface, a copolymer is formed which plays the role of a compatibi-
lizer, i.e. one is dealing with an *in situ* compatibilization phenomenon (Fig. 5.15(*a*)).
When one stimulates the progress of chemical interactions drastic changes in the
MFC are produced as schematically visualized in Fig. 5.15(*b*). The MFC system
we wish to discuss here is the one based on PA6 (as the matrix) and PET (as the
reinforcing microfibrils).

The different stages of MFC manufacture schematically presented in Fig. 5.15(*a*)
are better illustrated using a SEM and the selective extraction of the matrix (PA6);
see Fig. 5.16. The PET microfibrils which play the role of reinforcing elements are
rather impressive (Fig. 5.16(*b*)). As a result of profound chemical reactions, the
microfibrils form aggregates involving the PA6 matrix (Fig. 5.16(*c*)).

In order to examine the relationship between the microhardness of the MFC
and those of its constituents (PET and PA6) including the morphological entities, the two constituents of the MFC were subjected to the same thermal and
mechanical treatments as the MFC and characterized after each step. Further, to
evaluate the microhardness of the reinforcing microfibrils the additivity law was
applied and the effect of crystal size on the structure formation was taken into
account.

The fact that both the neat components and their blends are relatively well
characterized with respect to their varying structures and morphologies as a result
of the applied mechanical and thermal treatments, permits us to follow the gradual
variation of microhardness as a function of structural parameters. In this way
one can obtain the H values for material components which are not accessible to
direct experimental determination. Furthermore, having the extrapolated values
for completely amorphous and fully crystalline homopolymers and starting from
a knowledge of the number of components (and/or phases) one can make use of the
additivity law (eq. (1.5)) to evaluate the mechanical properties of components which
cannot be isolated or do not exist as individual materials. A good example of this
are the PET microfibrils studied here (Fig. 5.16(*b*)).

By extrapolating the straight-lines dependence of H on the degree of crystallinity
one can obtain the H values of the completely amorphous ($H_a^{PET} \sim 128$ MPa;
$H_a^{PA} \sim 52$ MPa) and fully crystalline ($H_c^{PET} \sim 294$ MPa; $H_c^{PA} \sim 283$ MPa)
components, which for PA6 is of particular importance since it is not accessible in
the fully amorphous state for such measurements. This is how the microhardness of
a fully amorphous PA6 was first evaluated (Krumova *et al.*, 1998).

As mentioned above, application of the additivity law (eq. (1.5)) supposes a
knowledge of the number of the components (or phases) with given microhardnesses
and weight fractions. What is not explicitly given in this equation is the type and the
extent of mutual dispersion of the components as well as the quality of the adhesion
on the contact surface boundary between the components (phases). We wish to
stress here that this has an influence on the reliability of the H values derived from
the additivity law.

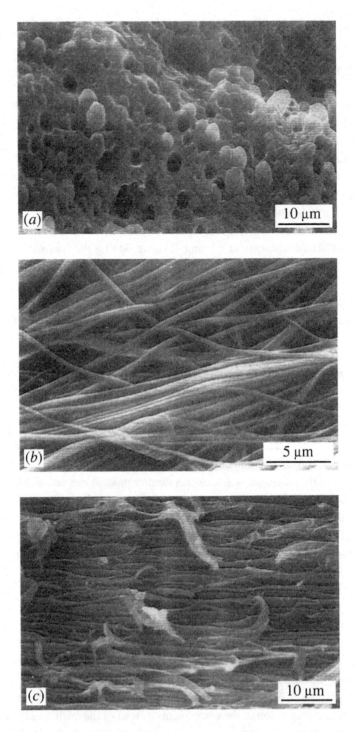

Figure 5.16. SEM micrographs of PET/PA6 blends: (*a*) as quenched (undrawn) taken from fracture surface after cooling in liquid N$_2$; (*b*) a blend drawn and annealed at 220 °C for 5 h after extraction of the PA6 component with a selective solvent; (*c*) a blend after cleavage of drawn and annealed fibres at 240 °C for 25 h (Krumova *et al.*, 1998).

In the present case we are dealing with well characterized samples with respect to the number of components and phases. The homopolymers consist of one component comprising two phases. Their blends are two-component with four different phases. Depending on the treatment conditions the number of components and/or of phases can be reduced. The conclusion that during annealing of the blends at 240 °C for longer times intensive chemical interactions take place is in accordance with many results on similar systems (blends of condensation polymers, Fakirov (1999)).

For this reason PET/PA6 blends treated at temperatures below 240 °C should be two-component and four-phase systems as is also shown from the WAXS

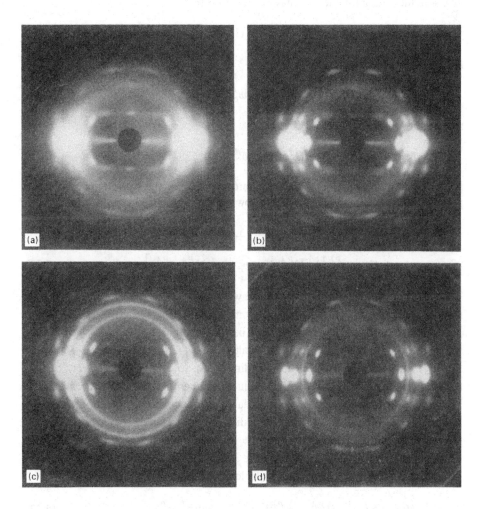

Figure 5.17. WAXS patterns of PET/PA6 (50/50 wt%) blend, zone drawn and annealed at different temperatures corresponding to the various stages of MFC processing: (*a*) after drawing at 180 °C, $\lambda = 4.2$; and annealing: at (*b*) 220 °C for 5 h, (*c*) 240 °C for 5 h, (*d*) 240 °C for 25 h (Krumova *et al.*, 1998).

experiments (Fig. 5.17). After annealing at 240 °C, especially for 25 h (Fig. 5.17(d)), the system has been converted into a two-component one (matrix and reinforcing microfibrils) of three- or possibly two-phase systems (amorphous matrix comprising 25% of the total amount of PET and almost fully crystalline one-phase microfibrils). Starting from these considerations and making use of the extrapolated data for the H_a and H_c values of the two homopolymers the additivity law (eq. (1.5)) has been used to derive the H values for the blends and the MFC and to compare them with the experimentally measured ones. Only after such a verification of the additivity law can the evaluation of the microhardness of the PET microfibrils be carried out.

Using the parallel model we may write for PET:

$$H^{PET} = H_c^{PET} w_c^{PET} + H_a^{PET}(1 - w_c^{PET}) \tag{5.19}$$

For PA we may write:

$$H^{PA} = H_c^{PA} w_c^{PA} + H_a^{PA}(1 - w_c^{PA}) \tag{5.20}$$

and for the blend:

$$H = H^{PET} w^{PET} + H^{PA}(1 - w^{PET}) \tag{5.21}$$

By combining eqs. (5.19)–(5.21) the microhardness for the two-component, four-phase system can be rewritten in the following form:

$$H = w^{PET}\left[H_a^{PET}(1 - w_c^{PET}) + H_c^{PET} w_c^{PET})\right]$$
$$+ (1 - w^{PET})\left[H_a^{PA}(1 - w_c^{PA}) + H_c^{PA} w_c^{PA})\right] \tag{5.22}$$

Then applying the w_c(DSC) data for the weight of each phase, the values derived for the microhardness H of the blends by means of eq. (5.21) are obtained.

Comparison shows good agreement between both sets of values. The only striking deviation is for the sample annealed at 240 °C for 25 h (Figs. 5.16(c) and 5.17(d)) for which H calculated, according to eq. (5.20), is much lower. Such a result means that the assumed H_c^{PET} values assigned in this particular case to the PET microfibrils are too low. In accordance with above consideration about the morphology of this sample one has to modify eq. (5.20), taking into account that for the same sample the crystallinity $w_c^{PA} = 0$ (see Fig. 5.17(d)). Since in the blend PET/PA6 = 1:1 (by wt), $w = 0.5$ and the additivity law for estimation of the H of PET microfibrils (eq. (5.22)) takes the following form:

$$H = 0.5\left(w_a^{PA} H_a^{PA} + w_a^{PET} H_a^{PET} + w_c^{PET} H_c^{PET}\right) \tag{5.23}$$

The first two terms refer to the matrix of MFC which is just one single-component amorphous phase consisting of random PET–PA6 copolymers.

Equation (5.23) for the microhardness of the amorphous matrix of the MFC (Fig. 5.17(d)) is acceptable as it has been demonstrated (by Baltá Calleja et al., 1998) that the H values for completely amorphous copolymers (with random sequential order) obey the additivity law provided the H_a values for the respective homopolymers are used. In this way one obtains for the microhardness H_c^{PET} of the microfibrils a value of 360 MPa. This value is higher than any other one reported for PET crystallized by the usual methods and approaches the measured value for PET crystallized under high pressure, the latter being 400 MPa (Baltá Calleja et al., 1994).

This surprisingly high H value for the reinforcing PET microfibrils can be explained by taking into account the peculiarities in the structure of the MFC (Fakirov et al., 1992).

Zone drawing results in highly-orientated chains and crystallites (Fig. 5.17(a)). Subsequent annealing at 220 °C results in crystallite growth and ordered stacking of crystalline lamellae and an increased degree of crystallinity (Fig. 5.17(b)). Direct TEM (including work on PET, one of the homopolymers discussed here, Peterman & Gohil, 1979, 1980; Peterman et al., 1981; Peterman & Rieck, 1987) has shown that the crystallization of oriented systems begins typically with the formation of fine microfibrillar precursor crystallites with the fibrils parallel to the fibre axis. These systems then transform into stacks of lamellar crystals, the stacking axis being parallel to the original fibrous crystallites and the chain axis still lying along the original fibre axis. The stacks can be relatively long in the stacking direction and narrow transverse to it. One can thus think of such an entity as a microfibril (Schultz & Peterman, 1984).

The microfibrils should be almost completely crystalline as can be concluded from the following consideration. The measured value w_c(DSC) of 75% refers to the total amount of PET in the blend. The amorphous part (25%) of the PET is involved in copolymers with PA6 as indicated above, but this is not the case with the crystalline PET (microfibrils). For this reason and because of the outlined structure formation peculiarities the microfibrils should be of very high crystallinity. This explanation is also supported by another observation. As mentioned above, the H of PET crystallized under pressure is 400 MPa (Baltá Calleja et al., 1994). Detailed structural analysis of such samples shows that they are almost completely crystalline ($w_c = 90\%$) and consist of rather large crystals (crystalline lamellae around 10–15 nm). Only by having this similarity in the structural characteristics does one obtain extremely high H values (Baltá Calleja, et al., 1994).

In the present case, isotropization due to melting of the PA6 component is established at $T_a = 240$ °C (Fig. 5.17(c)), the fibrillized PET preserving, however, its orientation and microfibrillar structure (Figs. 5.16(b) and 5.17(c), (d)). As a result a composite-like material is obtained comprising an isotropic semicrystalline (Fig. 5.17(c)) or non-crystalline (Fig. 5.17(d)) matrix of PA6 reinforced with almost fully crystalline, microfibrillized PET (Figs. 5.16 and 5.17).

In conclusion, annealing of a drawn PET/PA6 blend at 240 °C (i.e. between the T_m values of the two components) results in composites with a PA6-dominant amorphous matrix reinforced by the preserved PET microfibrils. In addition, the transreactions between the PET and PA6 lead to an improvement in the adhesion between matrix and microfibrils, resulting in high microhardness values. Furthermore, application of the additivity law makes possible the microhardness characterization of the fundamental elements of microfibrilar-reinforced composites – the microfibrils – which are not accessible for direct measurements. The surprisingly high value of 360 MPa obtained for the PET microfibrils is related to the peculiar structure of these morphological elements: the high degree of orientation and large crystallinity values.

5.7 References

Andresen, E. & Zachmann, H.G. (1994) *Colloid Polym. Sci.* **272**, 1352.

Andrews, E.H. (1974) *Pure Appl. Chem.* **39**, 179.

Apostolov, A.A., Boneva, D., Baltá Calleja, F.J., Krumova, M. & Fakirov, S. (1998) *J. Macromol. Sci. Phys.* **B37**, 543.

Arridge, R.G.C. (1975) *Mechanics of Polymers*, Clarendon Press, Oxford, p. 98.

Baltá Calleja, F.J. (1976) *Colloid Polym. Sci.* **254**, 258.

Baltá Calleja, F.J. (1985) *Adv. Polym. Sci.* **66**, 117.

Baltá Calleja, F.J. (1994) *Trends Polym. Sci.* **2**, 419.

Baltá Calleja, F.J. & Kilian, H.G. (1985) *Colloid Polym. Sci.* **263**, 697.

Baltá Calleja, F.J. & Kilian, H.G. (1988) *Colloid Polym. Sci.* **266**, 29.

Baltá Calleja, F.J., Fakirov, S., Roslaniec, Z., Krumova, M., Ezquerra, T.A. & Rueda, D.R. (1998) *J. Macromol. Sci. Phys.* **B37**, 219.

Baltá Calleja, F.J., Giri, L., Asano, T., Mieno, T., Sakurai, A., Ohnuma, M. & Sawatari, C. (1996) *J. Mater. Sci.* **31**, 5153.

Baltá Calleja, F.J., Giri, L., Ezquerra, T.A., Fakirov, S. & Roslaniec, Z. (1997a), *J. Macromol. Sci. Phys.* **B36**, 655.

Baltá Calleja, F.J., Giri, L. & Ward, I.M. (1995) *J. Mater. Sci.* **30**, 1139.

Baltá Calleja, F.J., Giri, L. & Zachmann, H.G. (1997b) *J. Mater. Sci.* **32**, 1117.

Baltá Calleja, F.J., Martínez-Salazar, J. & Asano, T. (1988) *J. Mater. Sci. Lett.* **7**, 165.

Baltá Calleja, F.J., Martínez-Salazar, J., Cackovic, H. & Loboda-Cackovic, J. (1981) *J. Mater. Sci.* **16**, 739.

Baltá Calleja, F.J., Ohm, O. & Bayer, R.K. (1994) *Polymer* **35**, 4775.

Baltá Calleja, F.J., Santa Cruz, C., Bayer, R.K. & Kilian, H.G. (1990a) *Colloid Polym. Sci.* **268**, 1.

Baltá Calleja, F.J., Santa Cruz, C., Chen, D. & Zachmann, H.G. (1991) *Polymer* **32**, 2252.

Baltá Calleja, F.J., Santa Cruz, C., Sawatari, C. & Asano, T. (1990b) *Macromolecules* **23**, 5352.

Buchner, S., Chen, D., Gehrke, R. & Zachmann, H.G. (1988) *Molec. Cryst. Liq. Cryst.* **155**, 357.

Denchev, Z., Sarkissova, M., Fakirov, S. & Yilmas, F. (1996) *Makromol. Chem.* **197**, 2869.

Denchev, Z., Sarkissova, M., Radusch, H.-J., Luepke, T. & Fakirov, S. (1998) *Makromol. Chem. Phys.* **199**, 215.

Devaux, J., Godard, P. & Mercier, J.P. (1982) *J. Polym. Sci. Polym. Phys. Ed.* **20**, 1875.

Ehring, R.J. (1992) (ed.) *Plastics Recycling*, Hanser, Munich, p. 93.

Evstatiev, M. & Fakirov, S. (1992) *Polymer* **33**, 877.

Fakirov, S. (ed.) (1999) *Transreactions in Condensation Polymers*, Wiley-VCH, Weinheim.

Fakirov, S. & Denchev, Z. (1999) in *Transreactions in Condensation Polymers* (S. Fakirov, ed.) Wiley-VCH, Weinheim, p. 319.

Fakirov, S. & Evstatiev, M. (1994) *Adv. Mater.* **6**, 395.

Fakirov, S., Apostolov, A.A., Boesecke, P. & Zachmann, H.G. (1990) *J. Macromol. Sci. Phys.* **B29**, 379.

Fakirov, S., Evstatiev, M. & Schultz, J.M. (1993) *Polymer* **34**, 4669.

Fakirov, S., Sarkissova, M. & Denchev, Z. (1996a) *Macromol. Chem.* **197**, 2837.

Fakirov, S., Sarkissova, M. & Denchev, Z. (1996b) *Macromol. Chem.* **197**, 2889.

Fernández-Berridi, M.J., Iruin, J.J. & Maiza, I. (1995) *Polymer* **36**, 1357.

Fitzer, E. (ed.) (1985) *Carbon Fibres and their Composites*, Springer-Verlag, Berlin.

Flores, A., Ania, F. & Baltá Calleja, F.J. (1997) *Polymer* **38**, 5447.

Fritz, H.G., Cai, Q., Bölz, U. & Anderlik, R. (1993) *IUPAC International Conference on Advanced Polymer Materials* (Jacobasch, H.J. ed.) Dresden 6–9 September, p. 5.

Fritz, H.G., Cai, Q., Cagiao, M.E., Giri, L. & Baltá Calleja, F.J. (1995) *J. Mater. Sci.* **30**, 3300.

Gibbs, M. (1990) *Plast. Eng.* **46**(7) 57.

Giri, L., Roslaniec, Z., Ezquerra, T.A. & Baltá Calleja, F.J. (1997) *J. Macromol. Sci. Phys.* **B36**, 335.

Gordon, M. & Taylor, J.S. (1952) *J. Appl. Chem.* **2**, 493.

Hull, D. (1981) *An Introduction to Composites Materials*, Cambridge University Press, Cambridge.

Karger-Kocsis, J. (1996) in *Polymeric Materials Encyclopedia* (Salamone, J.C. ed.) CRC Press, Inc., New York, p. 1378.

Kotliar, A.M. (1981) *J. Polym. Sci., Macromol. Rev.* **16**, 367.

Kimura, M., Salee, G. & Porter, R.S. (1984) *J. Appl. Polym. Sci.* **29**, 1629.

Krumova, M., Fakirov, S., Baltá Calleja, F.J. & Evstatiev, M. (1998) *J. Mater. Sci.* **33**, 2857.

Lambla, M., Yu, X. & Lorek, S. (1989) *Coreactive Polymer Alloys* ACS Symposia Series, American Chemical Society, Washington, p. 395.

Legge, N.R., Holden, G. & Schroeder, H.E. (eds.) (1987) *Thermoplastic Elastomers. A Comprehensive Review*, Hanser, Munich.

Lenz, R.W. & Go, S.S. (1973) *J. Polym. Sci., Polym. Chem. Ed.* **11**, 2927.

Lenz, R.W. & Go, S.S. (1974) *J. Polym. Sci., Polym. Chem. Ed.* **12**, 1.

Martínez-Salazar, J. & Baltá Calleja, F.J. (1985) *J. Mater. Sci. Lett.* **4**, 324.

Martínez-Salazar, J., Canalda Cámara, J.C. & Baltá Calleja, F.J. (1991) *J. Mater. Sci.* **26** 2579.

Martínez-Salazar, J., García Tijero, G.M. & Baltá Calleja, F.J. (1988) *J. Mater. Sci. Lett.* **23**, 862.

Matsuo, M., Inone, K. & Abumiya, N. (1984) *Sen'i Gakkaishi* **40**, 275.

Noland, J.S., Hsu, N.N.C., Saxon, R. & Schmitt, J.M. (1971) *Adv. Chem. Ser.* **99**, 15.

Paplham, W.P., Seferis, J.C., Baltá Calleja, F.J. & Zachmann, H.G. (1995) *Polym. Composites* **16**, 424.

Park, S.J., Kim, B.K. & Hong, H.M. (1990) *Eur. Polym. J.* **26**, 131.

Peterlin, A. (1973) *Pure Appl. Chem.* **8**, 277.

Petermann, J. & Gohil, R.M. (1979) *J. Mater. Sci.* **14**, 2260.

Petermann, J. & Gohil, R.M. (1980) *J. Polym. Sci., Polym. Lett. Ed.* **18**, 781.

Petermann, J. & Rieck, U. (1987) *J. Mater. Sci.* **22**, 1120.

Petermann, J., Gohil, R.M., Schultz, J.M., Hendricks, R.W. & Lin, J.S. (1981) *J. Mater. Sci.* **16**, 265.

Porter, R.S. & Wang, L.H. (1992) *Polymer* **33**, 2019.

Riedel, R. (1992) *Adv. Mater.* **4**, 759.

Roslaniec, Z., Ezquerra, T.A. & Baltá Calleja, F.J. (1995) *Colloid Polym. Sci.* **273**, 58.

Rueda, D.R., Baltá Calleja, F.J., Ayres de Campos, J.M. & Cagiao, M.E. (1988) *J. Mater. Sci.* **23**, 4487.

Rueda, D.R., Baltá Calleja, F.J., Viksne, A. & Malers, L. (1994) *J. Mater. Sci.* **29**, 1109.

Rueda, D.R., Cagiao, M.E., Baltá Calleja, F.J. & Palacios, J.M. (1987) *Synth. Met.* **22**, 53.

Santa Cruz, C., Baltá Calleja, F.J., Zachmann, H.G. & Chen, D. (1992) *J. Mater. Sci.* **27**, 2161.

Santa Cruz, C., Baltá Calleja, F.J., Zachmann, H.G., Stribeck, N. & Asano, T. (1991) *J. Polym. Sci., Polym. Phys. Ed.* **29**, 819.

Savage, G. (1993) *Carbon-Carbon Composites*, London: Chapman & Hall.

Sawatari, C., Shimogiri, S. & Matsuo, M. (1987) *Macromolecules* **20**, 1033.

Schultz, J.M. & Petermann (1984) *Colloid Polym. Sci.* **262**, 294.

Schroeder, H. & Cela, R. (1988) in *Encyclopedia of Polymer Science and Engineering*, Vol. 12 (Mark, M.F., Bikales, N.M., Overberger, C.G. & Menges, G. eds.) John Wiley & Sons, New York.

Skinner, J. & Gane, N. (1973) *Phil. Mag. A.* **28**, 827.

Sperling, L.H. (1986) *Introduction to Physical Polymer Science*, Wiley-Science, New York.

Stribeck, N., Sapundjieva, D., Denchev, Z., Apostolov, A.A., Zachmann, H.G. & Fakirov, S. (1997) *Macromolecules* **30**, 1329.

Chapter 6

Microhardness of polymers under strain

6.1   Polymorphic transitions in crystalline polymers

When polymers crystallize from the melt or solution, the crystalline regions may exhibit various types of polymorphic modification, depending on the cooling rate, evaporation rate of solvent, temperature and other conditions. These crystal modifications differ in their molecular and crystal structures as well as in their physical properties. Many types of crystalline modifications have been reported (Tashiro & Tadokoro, 1987). (See also Chapter 4.)

Some polymorphic modifications can be converted from one to another by a change in temperature. Phase transitions can be also induced by an external stress field. Phase transitions under tensile stress can be observed in natural rubber when it orients and crystallizes under tension and reverts to its original amorphous state by relaxation (Mandelkern, 1964). Stress-induced transitions are also observed in some crystalline polymers, e.g. PBT (Jakeways *et al.*, 1975; Yokouchi *et al.*, 1976) and its block copolymers with poly(tetramethylene oxide) (PTMO) (Tashiro *et al.*, 1986), PEO (Takahashi *et al.*, 1973; Tashiro & Tadokoro, 1978), polyoxacyclobutane (Takahashi *et al.*, 1980), PA6 (Miyasaka & Ishikawa, 1968), PVF$_2$ (Lando *et al.*, 1966; Hasegawa *et al.*, 1972), polypivalolactone (Prud'homme & Marchessault, 1974), keratin (Astbury & Woods, 1933; Hearle *et al.*, 1971), and others. These stress-induced phase transitions are either reversible, i.e. the crystal structure reverts to the original structure on relaxation, or irreversible, i.e. the newly formed structure does not revert after relaxation. Examples of the former include PBT, PEO and keratin.

Two crystalline phases of PBT can appear under tension and relaxation: the so called β and α forms (Hall & Pass, 1976) (see Fig. 6.1). The main difference in the molecular structure between the α and the β forms lies in the methylene conformation:

$$-O-CH_2-CH_2-CH_2-CH_2-O-$$

α	G	G	T	$\overline{G}$	$\overline{G}$
β	T	T	T	T	T

X-ray diffraction patterns and infrared Raman spectra show specific changes through this $\alpha \rightleftharpoons \beta$ transition (Tashiro & Tadokoro, 1987; Jakeways et al., 1976; Ward & Wilding, 1977). The stress and strain dependence of the molar fraction of the β form, X_β, can be evaluated by quantitative analysis of the infrared spectra. The characteristic behaviour of this phase transition observed in a uniaxially oriented sample is as follows: X_β increases drastically above the critical stress f^*; X_β is almost linearly proportional to the strain; the transition is reversible; and the

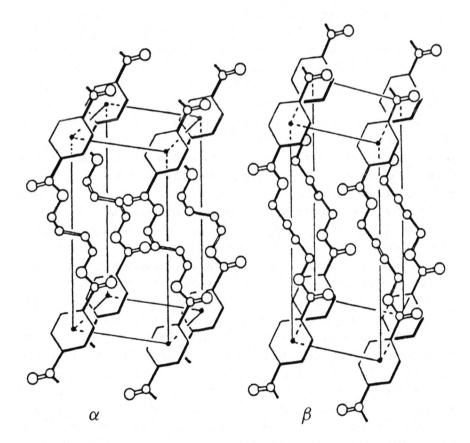

α β

Figure 6.1. Crystal structure of PBT: α and β forms. (From Tashiro & Tadokoro, 1987.)

stress–strain curve has a plateau at the critical stress f^*. Thus the curve is divided into three regions: the elastic deformation of the α phase (0–4% strain), the $\alpha \rightleftharpoons \beta$ transition (plateau region; 4–12% strain), and the elastic deformation of the β phase (>12% strain). These experimental results indicate that this stress-induced phase transition is a thermodynamic first-order transition.

It has been also pointed out that in block copolymers of PBT and PTMO, the transition is smeared over a wide range of stresses due to the effect of the PTMO soft segments (Tashiro *et al.*, 1986).

6.2   Stress-induced polymorphic transition in homopolymers, copolymers and blends

6.2.1   Effect of stress-induced polymorphic transition of PBT on microhardness

PBT or poly(tetramethylene terephthalate) became the subject of much interest in the mid 1970s when three groups of workers (Boyle & Overton, 1974; Jakeways *et al.*, 1975; Yokouchi *et al.*, 1976) independently reported that PBT could crystallize in two distinct polymorphic forms. The α form, as mentioned in the previous section, was found in a relaxed sample, whereas the β form could only be observed when the sample was held under strain. There have been a number of attempts to determine the unit cell parameters for the two crystalline forms (Desborogh & Hall, 1977; Stambaugh *et al.*, 1979; Mencik, 1975), and there is still some degree of controversy. However, the general consensus is that in the α form the molecular chain is not fully extended, probably with the glycol residue in a *gauche–trans–gauche* conformation, whereas in the β form the chain is fully extended with the glycol residue in the all-*trans* conformation. There has been considerable interest in the mechanism of the α–β transition, and this has been modelled for static and dynamic measurements (Brereton *et al.*, 1978; Davies *et al.*, 1980).

As we have seen in Chapter 4, microindentation hardness measurement provides a rapid evaluation of variations in surface mechanical properties of polymers affected by changes in microstructure including polymorphic transitions. In polymers such as PE, which display a lamellar morphology of stiff flat crystals intercalated by 'amorphous' compliant layers, the microhardness can be described by the simple additivity model expressed by eq. (4.3). As shown in Section 4.2.3, crystal microhardness, H_c, depends on the average thickness, ℓ_c, of the crystalline lamellae. However, in systems in which w_c and ℓ_c remain constant, H_c is a function of chain packing within the crystalline phase. Crystal microhardness – the critical stress required to plastically deform the crystal – hence reflects the crystal's response to the intermolecular forces holding the chains within the lattice and is a function of cohesion energy (Martínez-Salazar *et al.*, 1985).

In Section 4.5 we demonstrated that microhardness measurement is a technique capable of detecting polymorphic changes in polymers. In particular, the study of the transition from the α to the β form in i-PP confirmed that the changes in H can be explained in terms of an additive contribution to the H of independent phase components H_c^{α}, H_c^{β} and H_a (Baltá Calleja et al. 1988). This approach opens up the possibility of characterizing i-PP samples consisting of a mixture of α and β phases by means of H measurements.

Thus, it seems to be of interest to examine the influence of stress-induced polymorphic changes on the microhardness. While in the case of i-PP two samples comprising the α or β phase were characterized, here we wish to follow the micro-hardness behaviour during the α–β polymorphic transition caused by a mechanical field. For this purpose PBT has been selected as a suitable material because of its ability to undergo stress-induced polymorphic transition from the α (relaxed) to the β (strained) form. Bristles of commercial PBT with a diameter of about 1 mm were drawn at room temperature via neck formation (final diameter about 0.5 mm and draw ratio of 3.4) and thereafter annealed in vacuum at 200 °C for 6 h with fixed ends (Fakirov et al., 1998).

Although the stress-induced α–β polymorphic transition in PBT is well documented, comparative WAXS measurements of the samples were carried out in the same deformation range at which the H measurement are performed. In addition, the size of the coherently diffracting domains (crystal size) D_{hkl} in the (100) and (010) and ($\bar{1}$04) ('c' axis direction) during stretching was calculated from the integral breadth of the equatorial reflections according to $\delta\beta \sim 1/D_{hkl}$ (Baltá Calleja & Vonk, 1989).

H measurements of PBT under strains of up to 20% relative deformation (ε) were performed using a stretching device. The strain ε is defined as $\varepsilon = (\ell - \ell_0)/\ell_0$ where ℓ_0 and ℓ are the starting and stretched lengths of the sample, respectively. The indentation anisotropy $\Delta H = 1 - (d_{\parallel}/d_{\perp})^2$ was also derived (see eq. (2.6)). In order to evaluate the contribution of each polymorphic phase to the total H it is necessary to know their mass fractions at the different deformation stages as required by the additivity law (eq. (4.3)). For this purpose the data of Tashiro et al. (1980) obtained by the infrared study of the α–β transition have been used. The same authors have shown that just in the transition deformation interval ($\varepsilon = 4$–16%) the relationship between ε and the amount of the β phase is linear.

The values obtained are summarized in Table 6.1. The strain variation as a function of the H value is also presented in Fig. 6.2. One can see a very well defined decrease in H (from 150 to 120 MPa) in a rather narrow deformation interval (ε between 5 and 8%). Before this drop in H in the deformation range of $\varepsilon = 0$–5%, the microhardness is relatively constant at about 150 MPa, which is typical for semicrystalline PBT (Giri et al., 1997). After reaching its lowest value ($H = 118$ MPa) around $\varepsilon = 10\%$, H starts to increase almost linearly up

Table 6.1. *Microhardness perpendicular to the chain orientation (strain) direction* $H_\perp$, *indentation anisotropy at room temperature* (25 °C) ΔH^{25}, *X-ray crystallinity* w_c(WAXS), *percentage of* α *and* β *modifications (according to Tashiro et al., 1980) and crystal hardness of homo-PBT* H_c^β *stretched at various strains,* ε.

Property	Strain, ε (%)										
	0	2	4	6	8	10	12	14	16	18	20
$H_\perp$ (MPa)	152.9	144.8	156.0	144.4	126.6	118.2	120.2	125.7	133.7	144.4	193.5
ΔH^{25}	−0.0338	0.0071	0.0656	0.1603	0.2026	0.3799	0.2461	0.4059	0.3796	0.2460	0.4054
w_c(WAXS) (%)	47	47	48	48	49	50	50	50	49	47	46
α phase (%)	43.0	40.3	38.9	34.2	30.5	26.6	21.6	17.9	16.3		
β phase (%)	4.0	6.7	9.1	13.8	18.5	23.4	28.4	32.1	32.7		
H_c^β (MPa)	193.7	137.6	179.2	155.3	93.1	100.6	122.4	147.3	181.5		

to 190 MPa with the further rise in deformation up to $\varepsilon = 20\%$ when the sample breaks.

Figure 6.3 shows the WAXS radial scans of the ($\bar{1}04$) reflection of PBT taken at different stages of tensile deformation. One can see that the angular peak positions for the relaxed and slightly deformed ($\varepsilon = 5.3\%$) samples are the same ($2\theta = 31.5°$) in contrast to the samples deformed at higher strains ($\varepsilon = 12.9$ and 21.2%) for which the maximum is shifted to lower angles ($2\theta = 28°$). The observed change in the peak position is in agreement with earlier reports (Boyle & Overton, 1974; Jakeways & Wilding, 1975; Yokouchi & Sakakibara, 1976) and demonstrates that in the deformation interval between $\varepsilon = 5\%$ and $\varepsilon = 12\%$ a stress-induced polymorphic transition takes place.

The fact that the abrupt drop in H (Fig. 6.2) coincides with the well documented α–β transition in the deformation interval between 4 and 10-12% suggests that the observed H changes (Fig. 6.2) are exclusively related with the stress-induced polymorphic transition. Thus, the starting value of $H = 155$ MPa should be typical for PBT containing the crystalline α phase and the lower value of $H = 118$ MPa can be assigned to PBT comprising mostly the crystalline β phase. The observation that $H^{\alpha} > H^{\beta}$ is obviously related to the fact that the α modification is distinguished by

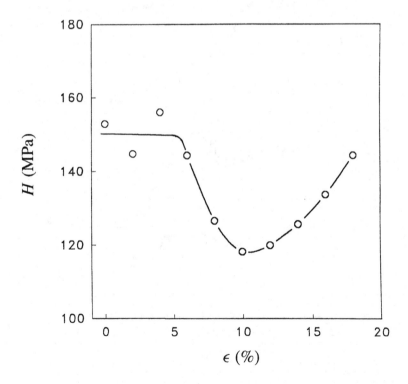

Figure 6.2. Effect of deformation ε on the microhardness H of homo-PBT. (From Fakirov & Boneva, 1998.)

a denser packing of chains as can be concluded from the ideal crystal densities. According to a comparison of the published crystalline polymorphic structures (Desborough & Hall, 1977), the volume of the unit cell for the α phase varies between 260.0 and 262.8 $\mathring{A}^3$ and that of β phase between 268.9 and 285.0 $\mathring{A}^3$. As mentioned above, it has been demonstrated that in cases in which w_c and ℓ_c do not change, the crystal microhardness H_c depends exclusively on the chain packing within the crystalline phase. In this way one can explain why $H^\alpha > H^\beta$, although other factors acting in the same direction can also exist as will be demonstrated below.

The strong increase in H obtained after completing the stress-induced polymorphic transition (from 118 to 150 MPa) can be explained by the additional chain

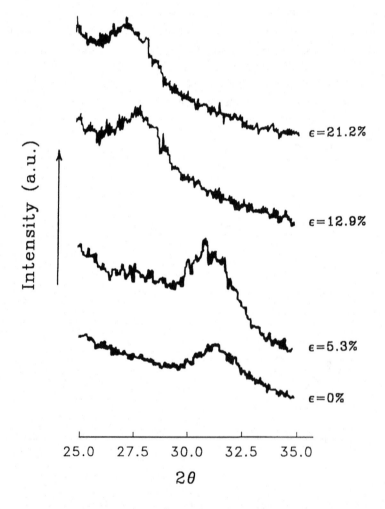

Figure 6.3. Radial WAXS scans of the ($\bar{1}$04) reflection of drawn annealed homo-PBT bristles taken at various tensile deformations ε. (From Fakirov *et al.*, 1998.)

orientation in the amorphous regions which results in an additional densification of the total structure. In fact, this densification effect should take place during the earlier stages of deformation where the indentation anisotropy also increases (Table 6.1). However, in this case the densification effect is compensated by the stress-induced polymorphic transition effect on H. Furthermore, it is hardly to be expected that the observed increase in H after the polymorphic transition (Fig. 6.2) is due to an increase of crystallinity during stretching since such a change cannot be found in the diffractograms of the WAXS (Table 6.1).

Let us return to the most interesting part of Fig. 6.2 – the deformation range in which the α–β transition occurs. The experimentally measured values of H allow one to calculate H^β provided the mass fraction of the crystallites of the β-type is known. For this purpose let us present eq. (4.3) in the following form:

$$H = H_c^\alpha \, w_c^\alpha + H_c^\beta \, w_c^\beta + \left[1 - (w_c^\alpha + w_c^\beta)\right] H_a \qquad (6.1)$$

where $w_c^\alpha + w_c^\beta \, (= w_c)$ are the mass fractions of α and β crystallites, respectively. Making use of the observation of Tashiro & Nakai (1980) that in the transition interval a linear relationship exists between the deformation ε and the amount of β phase obtained as a result of α–β transition, we can evaluate the w_c^β at each deformation stage in the deformation interval 4–16% (Table 6.1). For the microhardness of the amorphous phase the value $H_a = 54$ MPa was taken (Giri et al., 1997). The H_c^α values for each strain were defined for the corresponding crystal thickness using eq. (4.8), where $H_c^0 = 369$ MPa is the microhardness of infinitely large PBT crystals and $b = 15$ Å, as found by Giri et al. (1997).

Assuming that the amorphous phase does not crystallize further during this deformation interval, as is done for similar analysis by Tashiro & Nakai (1980) and is proven by WAXS in the present case (as can be concluded from the values of w_c in Table 6.1), one obtains H_c^β values of between 100 and 190 MPa.

The values of H_c^β values obtained can be grouped into three ranges (Table 6.1): (a) the deformation range before stress-induced polymorphic transition begins, $H_c^\beta = 150$–190 MPa; (b) the range in which the transition is completed ($\varepsilon = 10$–12%), $H_c^\beta \approx 110$ MPa; (c) the range of larger deformations after the completion of the transition, $H_c^\beta = 150$–180 MPa. One can assume that the most reliable values are those around 120 MPa for the following reasons. In the first group the amount of the β modification is relatively low (below 10%); the higher H_c^β values in the deformation range above $\varepsilon = 14\%$ (Fig. 6.2) cannot be related to the β phase. One can assume that the calculated increase in H_c^β in this deformation interval (Fig. 6.2) originates from the increase of the measured overall H due to the additional orientation of the amorphous phase giving rise to H_a. Such an assumption is supported by the systematic change in the indentation anisotropy ΔH, as can be concluded from Table 6.1. For this reason one can accept for H_c^β a value of 122 MPa. In fact, this value corresponds to the microhardness of fully crystalline PBT comprising crystallites of only the β-type. For comparison, recall that for the α phase this amounts to $H_c^\alpha = 287$ MPa as reported by Giri *et al.* (1977), which is considerably higher than H_c^β found in the present measurements. What could be the reason for this relatively large difference in the microhardness of the two polymorphic phases? In addition to the above mentioned difference in the unit cell volumes leading to lower H_c^β, there are at least other two factors which act in the same direction. As outlined above it should be stressed that the packing density will be the only H determining factor if the crystal size – as well as the crystal mass fraction – remain constant. This is not the case in the present study, at least with respect to crystal sizes, as can be seen from the data presented in Table 6.2. A significant crystal size decrease (up to 2–3 times) in the (100) and (010) directions has been found. This observation is in a very good agreement with the results of Roebuck *et al.* (1992). Also, in addition to the smaller crystal sizes found after the polymorphic transition, the above authors concluded that in order for such small crystals to be stable they had to be poorly connected to the bulk polymer. Both these outlined peculiarities of the β phase contribute to a decrease of H. In order to estimate the contributions of crystal size to H it is convenient to recall that the measurements of Krumova *et al.* (1998) on PA6 showed that a decrease in crystal size from 106 Å to 33 Å results in a reduction in the overall microhardness by 20%, keeping the crystalline mass fraction constant. (For more details about the dependence of H on crystal size see Section 4.3.)

One can conclude that the microindentation technique allows the strain-induced polymorphic transition in PBT to be followed. The observed rather abrupt variation in H (within 2–4% of external deformation) makes the method competitive with respect to sensitivity to other commonly used techniques such as WAXS, infrared spectroscopy, Raman spectroscopy, etc. (Tashiro & Tadokoro, 1987). Furthermore, by applying the additivity law it is possible to calculate the microhardness of completely crystalline PBT, comprising crystallites of the β-type, as $H_c^\beta = 122$ MPa. This technique can also be used to examine the stress-induced polymorphic behaviour of PBT in copolymers and blends as will be demonstrated in the following sections.

Table 6.2. *Crystal size D_{hkl}, of homo-PBT stretched at various strains ε.*

Crystal size (Å)	Strain, ε (%)			
	0	5.3	12.9	21.2
$D_{(100)}$	84.1	51.9	36.9	36.8
$D_{(010)}$	132.5	109.3	43.7	37.2
$D_{(\bar{1}04)}$	40.8	44.9	40.75	49.5

6.2.2   Microhardness behaviour during stress-induced polymorphic transition in block copolymers of PBT

From the preceding chapter one can conclude that thermoplastic elastomers are a special type of block copolymer exhibiting an extraordinary combination of reprocessability, elasticity, toughness, low-temperature flexibility and strength at relatively high temperatures (frequently *ca* 150 °C) (Legge *et al.*, 1987; Schroeder & Cella, 1988). For this reason, they are nowadays of great commercial importance. As mentioned above, the set of unique properties that they exhibit is mainly due to the existence of physical (temporary) network of cross-links tying the polymer chains into a three-dimensional network.

In a systematic study of PEE thermoplastic elastomers based on PBT as the hard segments and PEG as the soft segments an investigation of the deformation mechanism has been carried out (Stribeck *et al.*, 1997). For this purpose the relationship between the external (macro)deformation ε and the microdeformation (at a morphological level expressed by the changes in the long spacing L) was followed within a wide deformation range (between 0 and 200%) by means of SAXS (Fakirov *et al.*, 1991, 1992, 1993; Apostolov & Fakirov, 1992; Stribeck *et al.*, 1994, 1997). The affine and reversible increase in L at relatively low macrodeformation (up to $\varepsilon \approx 50\%$) is found to be related to reversible conformational changes in

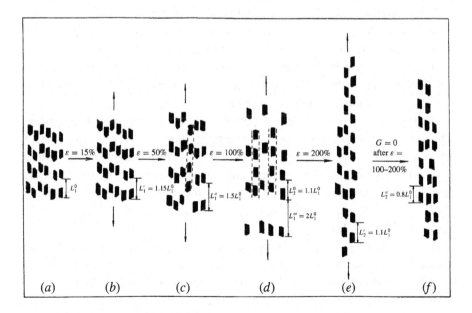

Figure 6.4. Schematic model of the structural changes in drawn semicrystalline thermoplastic elastomers of the PEE type at different stages of deformation: (*a*) no stress applied, (*b*)–(*e*) under increasing stress; (*f*) after stress removal (only crystalline regions are depicted). (From Fakirov *et al.*, 1991.)

the intercrystalline amorphous regions, in accordance with previous reports on PE (Gerassinov et al., 1978, 1979) and other thermoplastic elastomers (Legge et al., 1987; Pakula et al., 1985a,b). At this level of deformation there is no indication (from SAXS measurements) of any changes in the crystallites of PBT hard segments. The structural changes during deformation of PEE are shown schematically in Fig. 6.4. The model in Fig. 6.4(a) shows the structure of the starting material characterized by a long spacing L_1^0. The deformation of the sample within the range $\varepsilon = 10$–20% leads to the extension of the chains in the amorphous regions. The expansion of these regions results in a decrease of their density, hence giving rise to an increase in the scattering intensity and in the long period L by the same percentage. Since the inter- and intrafibrilar contacts remain unaffected by these deformations, it is expected that the changes described should be reversible, as is actually observed.

Further deformation of the sample up to $\varepsilon = 50\%$ (Fig. 6.4(c)) again results in an affine increase of the long spacing accompanied by a slight decrease of scattering intensity. The affine increase in L can, once more, be explained in terms of an extension of the chains within the amorphous regions. For this high deformation ($\varepsilon = 50\%$) a few microfibrils lose contacts with adjacent ones and relax. Such a relaxation causes a rise in the amorphous density and subsequently a decrease in scattering intensity as are actually observed.

Let us now discuss the microhardness of PBT block copolymers during the stress-induced polymorphic transition; for some block copolymers this transition is rather smeared (Tashiro et al., 1986). An additional reason for performing this work is the fact that the copolymers of PBT with PEO had not then been studied with reference to polymorphic transitions. The very detailed structural characterization of the copolymers with PEO as mentioned above was expected to shed more light on the nature of the microhardness behaviour.

The starting material represents a polyblock PEE comprising PBT as the hard segments and PEO (with a molecular weight of 1000 and polydispersity of 1.3, according to GPC analysis) (Fakirov et al., 1991) as the soft segments in a ratio of 57/43 wt%. The sample was a bristle, drawn at room temperature to five times its initial length, and then annealed with fixed ends for 6 h in vacuum at a temperature of 170 °C. WAXS and microhardness measurements were performed in the same way as for the homo-PBT.

The microhardness, degree of crystallinity and the percentage of α and β phase obtained are summarized in Table 6.3. The dependence of the microhardness H on the deformation ε for drawn and annealed bristles of PEE (PBT/PEO $= 57/43$ wt%) is plotted in Fig. 6.5. One can see that the H variation can be split into several regimes depending on the stress applied. For the lowest deformations (ε up to 25%) H is nearly constant ($\sim$33 MPa) and thereafter in a very narrow deformation interval ($\varepsilon = 2$–3%) H suddenly drops by 30% reaching the value of 24 MPa which is maintained in the ε range between 30 and 40%. With further increases in ε

Table 6.3. *Microhardness H, X-ray crystallinity w_c(WAXS), percentage of α and β modifications (according to Tashiro et al., 1980) of PEE stretched at various strains ε.*

Property					Strain, ε (%)														
	0	5	10	12	15	20	25	30	35	40	45	50	55	60	65	70	80
H_α (MPa)	34.2	26.6	32.7		32.4	33.8	34.2	24.2	24.2	27.0	29.1	29.5	32.7	33.5	28.7	30.2	15.8
w_c(WAXS) (%)	37	32	33	29				25						22			
α phase (%)	33.9	27.9	28.1	24.3				15.55						7.3			
β phase (%)	3.1	4.1	4.9	4.7				9.45						14.7			

(between 40 and 60%) H increases up to nearly its starting value, followed by decrease again to 25 MPa for $\varepsilon \sim$ 60–80%.

The most striking change in the H behaviour is the sharp decrease at $\varepsilon = $ 25–27%. Taking into account the similar study on homo-PBT (Fig. 6.2) where a similar change was observed (a drop in H by 20%) due to the well documented stress-induced polymorphic transition in PBT, one can assume that in the present case the same transition also takes place.

The fact that the α–β polymorphic transition in PEE occurs at a much higher deformation (around $\varepsilon = $ 27%) (Fig. 6.5) in comparison to the homo-PBT (ε between 5 and 10%) (Fig. 6.2) can be explained by the peculiar behaviour of the thermoplastic elastomers.

As stated above, the deformation of PEE with varying PBT/PEO ratio in the range $\varepsilon = $ 25–50% is related to the conformational changes in the amorphous regions, the crystallites remaining untouched. The stretching and relaxation of these amorphous segments causes an affine and reversible change of the measured X-ray long spacing as presented schematically in Fig. 6.4. With a further increase of deformation (starting at $\varepsilon = $ 25–30% for PEE with compositions as in this case) the external load is transferred to the crystallites contributing to the observed polymorphic transition (Fig. 6.5).

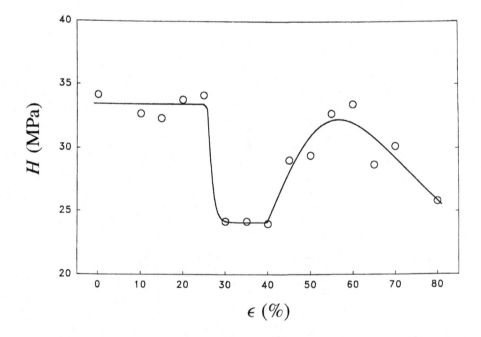

Figure 6.5. Microhardness H vs overall relative tensile deformation ε of drawn and annealed bristles of PEE (PBT/PEO = 57/43 wt%). (From Apostolov *et al.*, 1998.)

The drop in the microhardness is due to the fact that the β phase has a lower density than the α phase, as discussed in the previous section for the case of homo-PBT.

Additional support for the assumption concerning the stress-induced polymorphic transition around $\varepsilon = 27\%$ can be found in a WAXS study on the same samples for the same deformation interval. In conformity with the case of homo-PBT (Fig. 6.3), radial (near meridional) scans of the ($\bar{1}$04) reflection of PBT were taken from drawn and annealed PEE (PBT/PEO $= 57/43$ wt%) at various tensile deformations. It was found (Apostolov *et al.*, 1998) that up to $\varepsilon = 12\%$ the angular position of the ($\bar{1}$04) peak remains unchanged. However, for the diffractogram taken at $\varepsilon = 28.8\%$, in addition to the ($\bar{1}$04) diffraction peak there is another peak at $\sim 27°$. This is very well defined on the diffractogram taken at $\varepsilon = 58.8\%$ where the intensity of the new peak is higher than that of the starting one. This observation on PEE is in accordance with previous WAXS studies of homo-PBT during the stress-induced polymorphic transition (Jakeways & Smith, 1976; Nakamae *et al.*, 1982; Roebnek & Jakeways, 1992).

Similar changes can be found on the equatorial WAXS scans of drawn and annealed PEE (PBT/PEO $= 57/43$ wt%) taken at various tensile deformations (see Fig. 6.6). Once again, drastic changes, both with respect to the peak shape and peak angular position, can be detected at $\varepsilon = 28.8$ and 58.8%. The 'relative crystallinity' w_c (for these drawn samples) from WAXS at various stages of sample deformation is presented in Table 6.3 since calorimetry or density measurements cannot be used for strained samples.

Let us return to Fig. 6.5 and follow the H variation as a function of ε. The observed H increase which results in the restoration of the initial microhardness cannot be related to a regeneration of the α phase, distinguished by a higher microhardness H^{α}, because the sample is under stress. In order to explain the observed H increase one has to remember a peculiarity of the system: PEE is a physical network in which the crystallites are the cross-links. In addition, the crystallites are embedded in a soft amorphous matrix, i.e. they are 'floating' in a relatively low-viscosity matrix. During the deformation the network will stretch further depending on the relative 'softnesses' of the various components of the system. The stretching of the amorphous segments restricts the mobility of the crystallites and makes the system harder. In other words, the decrease of H caused by replacing of crystallites of a higher H by others of lower H is compensated by this microhardening effect caused by stretching of the network.

In addition to the assumed microhardening of the non-crystalline phases, the crystallization of the PEO soft segments in PEE acts simultaneously in the same deformation range as can be concluded from the analysis of WAXS data (see Fig. 6.6). The scattering curves at $\varepsilon = 28.8\%$ and $\varepsilon = 58.8\%$ are different from those at lower deformations. Their shape indicates overlapping of two independent reflections from two different unit cells. The lower-angle one arises from the β

modification of PBT segments in PEE as a result of the polymorphic transition, as shown in earlier studies (Roebuck *et al.*, 1992; Fakirov *et al.*, 1998). The higher-angle one does not belong to PBT but to the strain-induced crystallization of the soft PEO segments, as proven by Takahashi *et al.* (1973). The zigzag modification of PEO, formed at high tensile deformation is expected to show a higher microhardness H_c, due to its higher crystal density in comparison to the H_c of the usual (7/2) helix modification of PEO, without tensile deformation (Takahashi & Tadokoro, 1973).

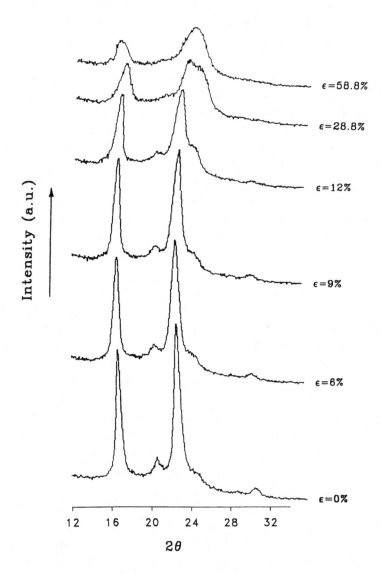

Figure 6.6. Equatorial WAXS scans of drawn and annealed bristles of PEE (PBT/PEO = 57/43 wt%) taken at various elongations. (From Apostolov *et al.*, 1998.)

The subsequent decrease of H in Fig. 6.5 back to 25 MPa again at the highest deformations ($\varepsilon = 60$–80%) can be explained by a partial destruction of crystallites due to the pulling out of hard segments from their crystallites. This mechanism has been demonstrated by SAXS and WAXS (Stribeck *et al.*, 1997) measurements. The introduction of defects in the crystallites in this way may induce a decrease in their H value.

A characteristic feature of the material under investigation is its very low microhardness – between 25 and 35 MPa depending on the crystalline modification present. These values are up to 5–6 times lower than those for semicrystalline homo-PBT regardless of the crystalline modification. Moreover, the obtained values for H of PEE (Table 6.3, Fig. 6.5) are about half the amorphous hardness, H_a, of PBT, being 54 MPa as reported by Giri *et al.* (1997). This means that there should be other factors responsible for the very low H values of the copolymer.

We have seen in Chapter 4 that microhardness is primarily determined by the crystalline phase, which in the case of polymers is always dispersed in an amorphous matrix. In the present case the PBT crystallites are embedded in a two-phase amorphous matrix, the amorphous PBT segments and the soft PEO segments as depicted on the model in Fig. 6.7 in which the four possible phases are shown. The PEO is distinguished by very low viscosity at room temperature – PEO of molecular weight of 1000 melts around 30 °C and being incorporated in a polymer chain its T_m is even lower – around 0 °C (Fakirov & Apostolov, 1990).

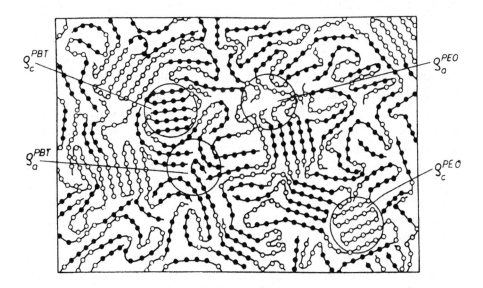

Figure 6.7. Model of the PEE block copolyester structure with hard (–●–●–●–) PBT segments and soft (–○–○–○–) PEO segments in equimolar amounts. Basically four phases are possible because both the PBT and PEO are crystallizable. (From Fakirov *et al.*, 1990.)

This means that during the first H decrease (see Fig. 6.5), PEO is in a molten state. It is worth mentioning that for the same reason the observed stress-induced polymorphic transition cannot arise from PEO, though it is known that PEO is capable of undergoing such a transition (Takahashi *et al.*, 1973; Tashiro & Tadokoro, 1978).

By means of the microhardness additivity law (eq. (4.3)) one can attempt to evaluate the contribution of the soft-segment amorphous phase to the overall micro-hardness. For a four-phase system as in the present case (Fig. 6.7), (two crystalline modifications and two amorphous phases), eq. (4.3) can be written as:

$$H = w\left[H_c^\alpha w_c^\alpha + H_c^\beta w_c^\beta + (1 - w_c^\alpha - w_c^\beta)H_a\right] + (1 - w)H^{soft} \quad (6.2)$$

where w is the mass fraction of the hard PBT segments, w_c^α and w_c^β are the degrees of crystallinity of the α and β modifications of PBT, respectively, H_c^α and H_c^β are their microhardnesses, $w(1 - w_c^\alpha - w_c^\beta)$ is the mass fraction of amorphous PBT with a microhardness H_a, and $(1 - w)$ is the mass fraction of the soft segments having a microhardness H^{soft}.

If we assume for H_c^β the value of 122 MPa which was described in the previous section on the stress-induced polymorphic transition in homo-PBT, then the calcu-lations of H, before the stress-induced polymorphic transition ($H = 34$ MPa) and after this transition ($H = 24$ MPa) would lead to negative values of H^{soft} of -112 and -70 MPa, respectively, which are of course not physically acceptable.

For this reason one has to revise the deformation mechanism during microhard-ness determination commonly used for complex systems comprising components or phases with glass transition temperatures below room temperature. For this purpose it is convenient to remember again the structural peculiarity of the system under investigation. The fact that the PBT crystallites are 'floating' in a matrix of low viscosity has important consequences on the microhardness behaviour. Because they are floating in a liquid of low viscosity, the crystallites of PBT as well as the PBT amorphous phase cannot respond to the external stress in a way that demonstrates their inherent microhardness, however, one can measure the response of the crystallites embedded in the liquid matrix.

A quite similar situation is observed when studying the microhardness behaviour of thermoplastic elastomers of PEEC-type comprising PBT as the hard segments (60 wt%) and PTMO and aliphatic polycarbonate (PCc) as the soft segments as already discussed in detail (see Section 5.4). In those cases the microhardness depression (i.e. the difference between the value calculated from eq. (6.2) and the measured value) $\Delta H = 35$–55 MPa was similar to the depression estimated in the present case and H_{soft} was evaluated to be between -90 and -140 MPa, depending on the amount of PCc in the soft phase (Baltá Calleja *et al.*, 1998).

Obviously in both cases the contribution of the soft-segment phase PEO or PTMO based on their overall mechanical resistance to external stress can be estimated via their viscosity. In contrast to the other soft amorphous solids the viscosity of these soft phases is not high enough to be characterized by means of microhardness (for which a residual indentation impression has to be formed). For this reason such low-viscosity liquids exhibit a microhardness decreasing effect. This effect is higher the lower the viscosity of the liquid-like phase (component) as demonstrated for various thermoplastic elastomers in Chapter 5.

Summarizing, it can be concluded that a relatively sharp (within 2–4% of deformation) drop in H is observed for copolymers of PBT but in comparison with homo-PBT this transition occurs at much higher deformations (between 25 and 30%). This difference as well as the following increase and decrease of H are related to the structural peculiarities of thermoplastic elastomers – the presence of a soft amorphous phase which first deforms and the existence of a physical network. The very low H values obtained for PEE are related to the fact that the PBT crystallites are 'floating' in an amorphous matrix characterized by a low viscosity.

6.2.3 Microhardness behaviour during stress-induced polymorphic transition in the blend of PBT with its block copolymers

In Sections 6.2.1 and 6.2.2 it has been demonstrated that the microhardness technique is very sensitive for detecting structural changes including polymorphic transitions in crystalline homopolymers and copolymers.

The study of the strain-induced polymorphic transitions by microhardness measurement offers the opportunity to gain additional information on the deformation behaviour of more complex polymer systems such as polymer blends. Since polymer blends are usually multicomponent and multiphase systems the question arises of how the independent components and phases react under the external load. The polymorphic transition will reflect the behaviour of the crystalline phases provided strain-induced polymorphic transition is possible.

After following the microhardness behaviour during the stress-induced polymorphic transition of homo-PBT and its multiblock copolymers attention is now focused on the deformation behaviour of a blend of PBT and a PEE thermoplastic elastomer, the latter being a copolymer of PBT and PEO. This system is attractive not only because the two polymers have the same crystallizable component but also because the copolymer, being an elastomer, strongly affects the mechanical properties of the blend. It should be mentioned that these blends have been well characterized by differential scanning calorimetry, SAXS, dynamic mechanical thermal analysis and static mechanical measurements (Apostolov et al., 1994).

The same PEE was used as in the preceding section (PBT/PEO = 49/51 wt%). For the preparation of the blend, both, the homopolymer PBT and PEE were cooled in liquid N_2 and finely ground. PBT was blended with PEE in the ratio PBT/PEE = 51/49 wt%. Bristles of the blend were prepared using a capillary rheometer, flushed with argon and heated to 250 °C. The melt obtained was kept in the rheometer for

5 min and then extruded through the capillary (diameter 1 mm). These bristles were annealed for 6 h at 170 °C in vacuum. The microhardness was measured at room temperature up to 30% overall relative deformation ε (at which fracture occurs) by using a stretching device. In accordance with the preceding measurements of microhardness under strain a deformation step of $\varepsilon = 5\%$ was also used in this case. The data obtained are plotted in Fig. 6.8.

Figure 6.8 illustrates the dependence of the microhardness on the external deformation for the PBT/PEE blend. One sees a sharp drop in H from 50 to 40 MPa at the very beginning of stretching (around $\varepsilon = 2$–3%). Based on previous studies (Sections 6.2.1 and 6.2.2) this sharp drop in H can be assigned to the $\alpha \rightarrow \beta$ polymorphic transition in PBT crystallites. Any further increase in deformation from $\varepsilon = 5\%$ to $\varepsilon = 25\%$ does not cause any pronounced changes in H. Such a constancy of H suggests that no transitions take place in this deformation interval. The next transition starts at $\varepsilon = 25\%$ and seems to be completed in a rather narrow deformation interval of $\varepsilon = 5\%$ (between $\varepsilon = 25$ and 30%). The occurrence of two distinct transitions, taking place in quite different deformation ranges ($\varepsilon = 5\%$ and

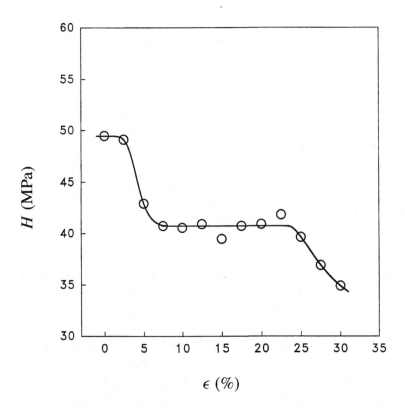

Figure 6.8. Variation of microhardness H with increasing relative deformation ε of the blend PBT/PEE = 49/51 wt% (with PEE of PBT/PEG-1000 = 51/49 wt%). (From Boneva *et al.*, 1998.)

$\varepsilon = 25\%$), suggests that the PBT crystallites present in the blend differ significantly in their response to the external mechanical load. DSC and X-ray studies show the existence of two types of PBT crystallites: those of the homo-PBT and those of the PBT segments from the polyblock PEE (Apostolov *et al.*, 1994; Gallagher *et al.*, 1993). One can assume that the first strain-induced polymorphic transition arises from the homo-PBT crystallites and the second one, appearing at higher deformation range, can be assigned to PBT crystallites belonging to PEE. This assumption is supported by the curves shown in Fig. 6.9 in which the results from Fig. 6.8 are replotted together with the data for homo-PBT (from Fig. 6.2) and its multiblock copolymer PEE (from Fig. 6.5).

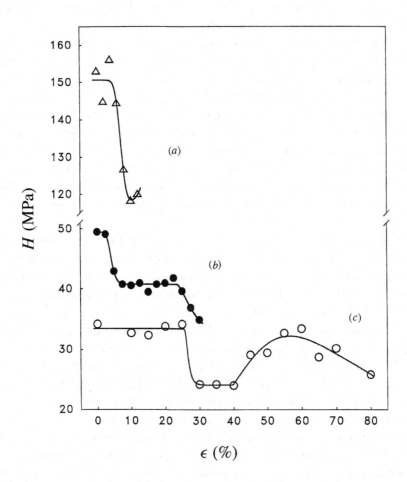

Figure 6.9. Variation of microhardness H with increasing relative deformation ε for: (*a*) homo-PBT (from Fig. 6.2), (*b*) the blend PBT/PEE = 49/51 wt% (with PEE of PBT/PEG-1000 = 51/49 wt% (from Fig. 6.8), and (*c*) the multiblock copolymer PEE with PBT/PEG = 57/43 wt% (from Fig. 6.5). For a better visual presentation a different H scale for samples (*b*) and (*c*) has been used. (From Boneva *et al.*, 1998.)

One can see here the fairly good agreement in the deformation range for the blend and the homo-PBT, on the one hand (Fig. 6.9(a) and (b)) and the blend and the copolymer on the other (Fig. 6.9(b) and (c)). In the first case the strain-induced polymorphic transition is observed at deformations around $\varepsilon = 5\%$ and in the second case at ε around 25%.

Comparison of the curves presented in Fig. 6.9 for homo-PBT (a), the blend (b) and the copolymer of PBT with PEG (c) allows one to draw the following conclusions: the initial drop in H at around $\varepsilon = 5\%$ (Fig. 6.8) originates from the strain-induced polymorphic transition in the crystallites comprising only homo-PBT segments. The second change in H at around $\varepsilon = 25-30\%$ is related to the strain-induced polymorphic transition in the PBT crystallites comprising PBT segments from its multiblock copolymer PEE.

Another striking observation from Fig. 6.9 is the fact that the numerical value for the experimentally measured H of the blend (Fig. 6.9(b)) is much lower than the one calculated according to the additivity law (eq. (4.3)) using the values for the homo-PBT and the copolymer PEE (Figs. 6.9(a) and (c), respectively). One possible explanation is that this may be due to the presence of two types of PBT crystallites which differ significantly in their perfection as demonstrated by Apostolov et al. (1994). Another possible explanation is that it may be related to the strong influence of the crystal surface free energy on H (Baltá Calleja, 1985). The latter would contribute to the decrease of the crystal microhardness particularly because of more defective PBT crystallites arising from the PEE copolymer (see also Section 4.4).

The above data may lead one to ask what is the origin of the different behaviours for the two types of crystallites with respect to their two-step response to the external mechanical field. Before answering this question let us recall some structural peculiarities of the system under investigation which are presented schematically in Fig. 6.10. (1) The blend contains the same crystallizable component in both, the homopolymer and the copolymer. In other words, the PBT crystallites can arise as a result of complete cocrystallization, i.e. formation of uniform crystallites with simultaneous participation of PBT from both the homopolymer and the copolymer as found for PEE containing 75–91 wt% PBT (Fig. 6.10(a)). (2) For reasons related to the length of the crystallizable blocks in the PEE copolymers, the more frequent case is that of partial cocrystallization, i.e. the formation of (continuous) crystals consisting of two spatially not separated, crystallographically identical populations of crystallites, differing in size, perfection, origin and time of appearance (Fig. 6.10(b)). This type of cocrystallization is observed when the PBT/PEE blends are drawn and thereafter annealed at a temperature between the melting temperatures of the two species of crystallites (Apostolov et al., 1994). (3) Crystallization with no cocrystallization can occur, i.e. formation of two types of PBT crystallites comprising either only segments from homo-PBT or only segments from PEE.

The microhardness data presented in Fig. 6.8 suggest that in the present case no cocrystallization occurs. If the PBT crystallites had resulted from complete cocrystallization (case (1)) one would expect one single strain-induced polymorphic transition in the whole deformation interval. If there had been partial cocrystallization (case (2)), one would observe a more or less continuous polymorphic transition between the deformation ranges typical for the homo-PBT and the PEE. However, the experimental results in Fig. 6.8 show two rather sharp transitions separated from each other by $\varepsilon = 20\%$. This finding supports the assumption that for the blend under investigation one is dealing with two crystallographically identical, but spatially separated, species of crystallites with different origin for the PBT segments (case (3)).

This conclusion about the lack of cocrystals in the PBT/PEE blend under consideration helps us to better understand the observed mechanical behaviour of the system.

The result shown in Fig. 6.8 that the two species of crystallites respond to the mechanical field in sequence – first the homo-PBT crystallites and later those arising from PEE, means that the homo-PBT crystals are probably dispersed within PEE in such way that they experience the mechanical field from the very beginning of loading. Moreover, one can assume that in the blend some internal stress and/or strain pre-exists since the strain-induced polymorphic transition starts even at lower

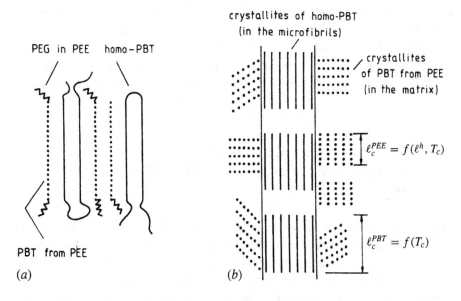

Figure 6.10. Schematic representation of the two types of cocrystallization of homo-PBT and PEE (with different PBT contents) in drawn blends of the two polymers: (a) complete cocrystallization (PBT in PEE is 75–91 wt%), (b) partial cocrystallization (PBT in PEE is below 75 wt%, in the present case it is 50 wt%). ℓ_c = lamella thickness, T_c = crystallization temperature, ℓ^h = length of the hard segments (PBT) in PEE. (From Apostolov et al., 1994.)

deformations (about $\varepsilon = 2$–3%, Fig. 6.8) than in the case of homo-PBT where $\varepsilon = 5\%$ for the same transition (Fig. 6.2).

After the well pronounced mechanical response of the homo-PBT crystallites, one observes conformation changes (stretching) only of the amorphous intercrystalline layers as concluded from SAXS measurements and discussed in Sections 6.21 and 6.2.2. These processes dominate up to deformations of about $\varepsilon = 25\%$ when the applied stress is also experienced by the crystallites arising from the PEE copolymer (Fig. 6.8).

Thus, the observation of two sharp well defined, and clearly separated on the deformation scale, strain-induced polymorphic transitions convincingly demonstrates that the two populations of PBT crystallites of differing origin, undergo the mechanical loading not simultaneously but in two steps: first those comprising homo-PBT (at $\varepsilon = 2$–3%), followed by crystallites belonging to PEE (at $\varepsilon = 25\%$).

In summarizing the results from the last three sections, one can conclude that the systematic variation of microhardness under strain performed on (*a*) homo-PBT (Section 6.2.1), (*b*) its multiblock copolymer PEE (Section 6.2.2) and (*c*) on blends of both of these (this section) is characterized by the ability of these systems to undergo a strain-induced polymorphic transition. The ability to accurately follow the strain-induced polymorphic transition even in complex systems such as polymer blends allows one also to draw conclusions about such basic phenomena as cocrystallization. In the present study of a PBT/PEE blend two distinct well separated (with respect to the deformation range) strain-induced polymorphic transitions arising from the two species of PBT crystallites are observed. From this observation it is concluded that: (i) homo-PBT and the PBT segments from the PEE copolymer crystallize separately, i.e. no cocrystallization takes place, and (ii) the two types of crystallites are not subjected to the external load simultaneously but in a sequential manner.

6.3   Reversible microhardness in polyblock thermoplastic elastomers with PBT as the hard segments

From the microhardness behaviour during the strain-induced polymorphic transition of PBT, differences were found for the three above described systems. The common characteristic feature among PBT (Section 6.2.1), its copolymer PEE (Section 6.2.2) and the PBT/PEE blend (Section 6.2.3) is the relatively sharp (within 2–4% of deformation) stepwise decrease in H (typically by 20–30% of starting H value). This drop appears at different deformation intervals: for PBT the sharp decrease occurs between 5–8% (Fig. 6.2), for PEE it appears between 25–30% (Fig. 6.5), while for the blends one observes two sharp decreases with increasing strain (Fig. 6.8). The first 20% decrease in the starting H value coincides with the deformation interval typical for the homo-PBT. The second decrease, amounting to another 10% drop in H, coincides with the interval observed for the PEE (Fig. 6.8). The observed extremely sharp change in H (within 2–4% of external overall deformation) makes the microhardness method competitive, as regards sensitivity, with other commonly used techniques such as WAXS, infrared and Raman spectroscopy, and other methods employed for the detection of the strain-induced polymorphic transitions in crystalline polymers.

Since thermoplastic elastomers can undergo large (up to a couple of hundred per cent) deformations which are to a great extent (up to 50%) reversible, we wish to discuss here the microhardness behaviour of these systems at different deformation levels, both, while maintained under stress, and after removal of the stress. In this way one may expect to observe reversibility in the microhardness values in so far as the polymer structure is restored after removing the stress.

As mentioned above, PEE, is well characterized mainly by SAXS in the deformation regime under consideration (Fakirov et al., 1991, 1992, 1993, 1994; Stribeck et al., 1997). In addition to these morphological investigations a study of the strain-induced polymorphic transition in PEE using microhardness measurement will shed additional light on the stress- and strain-induced structural reorganization of this new class of polymeric materials.

The same PEE material was used for this study as in the previous sections. Using a stretching device it was possible to perform measurements up to 70% overall relative deformation ε at which point the sample broke. Again the deformation was increased in steps of 5%. The main difference was that in previous studies (Sections 6.2.1–6.2.3) the deformation was increased continuously without relaxation, whereas in this case the sample was unloaded and allowed to relax after each H measurement under stress before the next H measurement without stress ($\sigma = 0$) was performed. The sample was then stretched to the next deformation and H was measured again. It should be noted that beyond some overall deformation (typically $\varepsilon > 20\%$) the unstressed sample shows some residual (plastic) deformation amounting about 50% of the strain ε under stress. The same deformation cycle has been used for SAXS measurements aimed at the morphological characterization of these PEE samples.

Figure 6.11(a) shows H as a function of the deformation under stress and Fig. 6.11(b) H measured after removing the tensile stress on the sample, i.e. the deformations in (b) (related ones) correspond to the residual, plastic deformations. In accordance with common practice (Baltá Calleja, 1985) H values in the direction perpendicular to the draw direction ($H_\perp$) have been used and are designated as just H. They correspond to the plastic deformation of the oriented polymer material under the indenter.

After the initial H increase, as a result of the first loading–unloading cycle ($\varepsilon = 5\%$), a drastic decrease of hardness (by 50%) is observed (Fig. 6.11(a)). With further increasing deformation under stress (in the range $\varepsilon = 20$–70%) H shows a very

slight increase, followed by a final decrease (Fig. 6.11(a)). The initial strong H
decrease in the deformation range 5–10% is much larger than the one observed for
homo-PBT (Fig. 6.2), for its copolymers (PEE) (Fig. 6.5) or blends (Fig. 6.8) where
it amounts to 20–30% of the starting H values. There is another important difference
between the present results and the previous ones (Figs. 6.2, 6.5 and 6.8). For PEE
(Fig. 6.2), in particular, the H decrease takes place at much lower deformation. This
is typical of homo-PBT (Fig. 6.2) but not of PEE where it occurs in the range $\varepsilon =$
25–30%. Such a difference in the microhardness behaviour of PEE could be related
to the rather different deformation regimes in these cases. While in the previous
cases deformation continuously increases maintaining the sample constantly under
stress, in the present case the PEE sample was allowed to relax ($\sigma = 0$) after each
deformation step ($\varepsilon = 5\%$).

It is worth noting that the H values for the homo-PBT (Fakirov et al., 1998)
are at least twice as high as for its copolymers (Apostolov et al., 1998) and blends
(Boneva et al., 1998). This observation is related to the specific role played by
the soft-segment phase during deformation. This subject is discussed in detail in
Sections 6.1 and 6.2 and is beyond the scope of the discussion here where interest

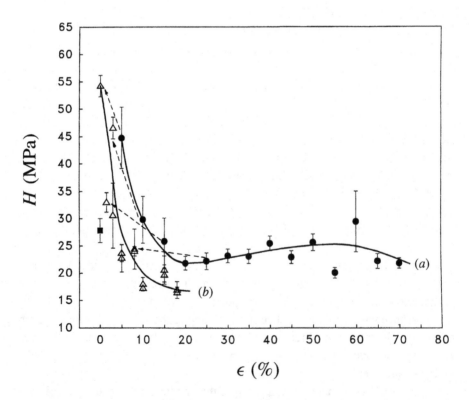

Figure 6.11. Dependence of microhardness H on strain ε: (a) measured under stress (●); (b)
measured after relaxation (△); starting value (■). (From Baltá Calleja et al., 1998.)

is focused on the relative difference between H_c^α and H_c^β and not on their absolute values; even more so because to the experimentally obtained H values contribute not only the two crystalline modifications but also the two amorphous phases arising from the hard (PBT) and soft (PEG) segments.

For the PEE under investigation it was shown that beyond a given deformation range, which depends on the ratio of the hard/soft segments, but typically is above $\varepsilon = 20$–25%, the observed overall deformation under stress is only partially reversible (usually about 50%) (Stribeck *et al.*, 1997). This means that the remaining deformation under stress is a consequence of a plastic deformation. In Fig. 6.11(b) the microhardness H is plotted as a function of the residual plastic deformation as discussed above.

One immediately sees that Fig. 6.11(b) has a completely different shape and position in comparison to that measured under stress (Fig. 6.11(a)). Indeed, a continuous decrease of the H values in the entire rather narrow remaining deformation interval is observed (plastic deformation between 0 and 20%). The most interesting observation here is that the H values for plastic deformations above 5% (corresponding to overall deformations ε between 25 and 70%) are very close to each other and small in magnitude (close to or even smaller than the lowest values measured under stress. In contrast to this group of H values, the ones corresponding to plastic deformations smaller than 5% are larger than the H values measured under tensile stress ($\sigma \neq 0$). The arrows in Fig. 6.11 indicate the sequence of measurement, i.e. after which deformation under stress the relaxed H values were taken.

In fact, the most interesting H data are measured in the relaxed state ($\sigma = 0$) for the lowest plastic deformation interval (Fig. 6.11(b)) where the most striking observation is that the H values for residual plastic deformations up to 5% are much higher than the rest of the H values. These H values are in fact even higher than those measured under strain (compare the H values under strain, $\varepsilon = 5$–15, Fig. 6.11(a) with those after relaxation, Fig. 6.11(b)). In contrast, when the residual plastic deformation exceeds 5% (for the steps with higher than 25% overall deformation) the H values are very low.

The occurrence of these two groups of values (below and above 5% residual deformation) can be explained by the strain-induced α–β polymorphic transition in PBT. As stressed above, it is well known (Yokouchi *et al.*, 1976) that up to 5% deformation the α polymorphic modification characterized by higher microhardness H_c^α, dominates in the samples. Furthermore, for 12–15% deformation (for homo-PBT), the α–β transition is essentially completed (see Fig. 6.11(a)) and the samples show predominantly the β polymorphic modification, which has a lower microhardness $H_c^\beta < H_c^\alpha$. However, after removal of the load ($\sigma = 0$) the samples contract (e.g. after a deformation of $\varepsilon = 5$–10%, the plastic deformation is around 1% and after $\varepsilon = 15$–20% the plastic deformation is around 3%). In all these cases the plastic deformation is less than 5%, i.e. the α modification dominates with its higher H values. This suggests that the increase in H observed after relaxation, as compared to the stressed cases (provided plastic deformation does not exceed 5%), is related to the regeneration of the initial α modification due to the reversibility of the $\alpha \Leftrightarrow \beta$ polymorphic transition.

The results shown in Fig. 6.11 not only support the reversibility of the strain-induced polymorphic transition (Boyle & Overton, 1974) but also allow one to speak about reversibility of microhardness. This is feasible in cases in which, as a result of some treatment, it is possible to regenerate the starting structure of the polymer. Reversibility of the microhardness further emphasizes that this mechanical property depends primarily on the structure of material.

In summary, it can be concluded that the microhardness technique is sensitive enough to detect strain-induced polymorphic transitions in polymers. The results in this chapter reveal that in materials characterized by a high and reversible deformation ability it is possible to observe reversible microhardness provided the strain-induced structural changes are reversible too.

6.4   References

Apostolov, A.A. & Fakirov, S. (1992) *J. Macromol. Sci. Phys.* **B31**, 329.

Apostolov, A.A., Boneva, D., Baltá Calleja, F.J., Krumova, M. & Fakirov, S. (1998) *J. Macromol. Sci. Phys.* **B37** 543.

Apostolov, A.A., Fakirov, C., Sezen, B., Bahar, I. & Kloczkowski, A. (1994) *Polymer* **35**, 5247.

Astbury, W.T. & Woods, H.J. (1933) *Phil. Trans. Roy. Soc. London A* **232**, 333.

Baltá Calleja, F.J. (1985) *Adv. Polym. Sci.* **66** 117.

Baltá Calleja, F.J. & Fakirov, S. (1997) *Trends Polym. Sci.* **5**, 246.

Baltá Calleja, F.J. & Vonk, C.G. (1989) *X-ray Scattering of Synthetic Polymers* (A.D. Jenkins ed.) Elsevier, Amsterdam.

Baltá Calleja, F.J., Fakirov, S., Roslaniec, Z., Krumova, M., Ezquerra, T.A. & Rueda, D.R. (1998) *J. Macromol. Sci. Phys.* **B37**, 219.

Baltá Calleja, F.J., Martínez-Salazar, J. & Asano, T. (1988) *J. Mater. Sci. Lett.* **7**, 165.

Boneva, D., Baltá Calleja, F.J., Fakirov, S., Apostolov, A.A. & Krumova, M. (1998) *J. Appl. Polym. Sci.* **69** 2271.

Boyle, C.A. & Overton, J.R. (1974) *Bull. Am. Phys. Soc.* **19** 352.

Brereton, M.G., Davies, G.R., Jakeways, R., Smith, T. & Ward, I.M. (1978) *Polymer* **19** 17.

Davies, G.R., Smith, T. & Ward, I.M. (1980) *Polymer* **21** 221.

Desborough, I.J. & Hall, I.H. (1977) *Polymer* **18**, 825.

Fakirov, S., Apostolov, A.A., Boeseke, P. & Zachmann, H.G. (1990) *J. Macromol. Sci. Phys.* **B29**, 379.

Fakirov, S., Boneva, D., Baltá Calleja, F.J., Krumova, M. & Apostolov, A.A. (1998) *J. Mater. Sci. Lett.* **17**, 453.

Fakirov, S., Denchev, Z., Apostolov, A.A., Stamm, M. & Fakirov, C. (1994) *Colloid Polym. Sci.* **272**, 1363.

Fakirov, S., Fakirov, C., Fischer, E.W. & Stamm, M. (1991) *Polymer* **32**, 1173.

Fakirov, S., Fakirov, C., Fischer, E.W. & Stamm, M. (1992) *Polymer* **33**, 3818.

Fakirov, S., Fakirov, C., Fischer, E.W., Stamm, M. & Apostolov, A.A. (1993) *Colloid Polym. Sci.* **271**, 881.

Gallagher, K.P., Zhang, X., Runt, J.P., Huynh-ba, G. & Lin, J.S. (1993) *Macromolecules* **26**, 588.

Gerasimov, V.I., Zanegin, V.D. & Smirov, V.D. (1979) *Visokomol. Soedin. Ser A*, **21**, 756 (in Russian).

Gerasimov, V.I., Zanegin, V.D. & Tsvankin, D.Ya. (1978) *Visokomol. Soedin. Ser A*, **20**, 846 (in Russian).

Giri, L., Roslaniec, Z., Ezquerra, T.A. & Baltá Calleja, F.J. (1997) *J. Macromol. Sci. Phys.* **B36** 335.

Hall, I.H. & Pass, M.G. (1976) *Polymer* **17**, 807.

Hasegawa, R., Takahashi, Y., Chatani, Y. & Tadokoro, H. (1972) *Polym. J.* **3**, 600.

Hearle, J.W.S., Chapman, B.M. & Senior, G.S. (1971) *Appl. Polym. Symp.* **18**, 775.

Jakeways, R., Smith, T., Ward, I.M. & Wilding, M.A. (1976) *J. Polym. Sci. Polym. Lett. Ed.* **14**, 41.

Jakeways, R., Ward, I.M., Wilding, M.A., Hall, I.H., Desborough, I.J. & Pass, M.G. (1975) *J. Polym. Sci. Polym. Phys. Ed.* **13**, 799.

Krumova, M., Fakirov, S., Baltá Calleja, F.J. & Evstatiev, M. (1998) *J. Mater. Sci.* **33** 2857.

Lando, J.B., Olf, H.G. & Peterlin, A. (1966) *J. Polym. Sci. Part A* **4**, 941.

Legge, R., Holden, G. & Schroeder, H. (eds.) (1987) *Thermoplastic Elastomers, Research and Development*, Car Hanser Verlag, Munich.

Mandelkern, L. (1964) *Crystallization of Polymers*, McGraw-Hill Inc, New York.

Martínez-Salazar, J., García, J. & Baltá Calleja, F.J. (1985) *Polym. Commun.* **26** 57.

Mencik, Z. (1975) *J. Polym. Sci. Polym. Phys. Ed.* **13** 2173.

Miyasaka, K. & Ishikawa, K. (1968) *J. Polym. Sci. Part A-2* **6**, 1317.

Nakamae, K., Kameyama, M., Yoshikawa, M. & Matsumoto, T. (1982) *J. Polym. Sci. Polym. Phys. Ed.* **20**, 319.

Pakula, T., Saijo, K. & Hashimoto, T. (1985a) *Macromolecules* **18** 1294.

Pakula, T., Saijo, K. & Hashimoto, T. (1985b) *Macromolecules* **18** 2037.

Prud'homme, R.E. & Marchessault, R.H. (1974) *Macromolecules* **7**, 541.

Roebuck, J., Jakeways, R. & Ward, I.M. (1992) *Polymer* **33** 227.

Schroeder, H. & Cella, R.G. (1988) in *Encyclopaedia of Polymer Science and Engineering*, Vol. 12 (Mark, H.F., Bikales, N.M., Overberger, C.G. & Menges, C. eds.) John Wiley & Sons, New York.

Stambaugh, B., Koenig, J.L. & Lando, J.B. (1979) *J. Polym. Sci. Polym. Phys. Ed.* **17** 1053.

Stribeck, N., Apostolov, A.A., Zachmann, H.G., Fakirov, C., Stamm, M. & Fakirov, S. (1994) *Int. J. Polym. Mater.* **25**, 185.

Stribeck, N., Sapundjieva, D., Denchev, Z., Apostolov, A.A., Zachmann, H.G., Stamm, M. & Fakirov, S. (1997) *Macromolecules* **30**, 1329.

Takahashi, Y. & Tadokoro, H (1973) *Macromolecules* **6**, 672.

Takahashi, Y., Osakai, Y. & Tadokoro, H. (1980) *J. Polym. Sci. Polym. Phys. Ed.* **18**, 1863.

Takahashi, Y., Sumita, I. & Tadokoro, H. (1973) *J. Polym. Sci. Polym. Phys. Ed.* **11**, 2113.

Tashiro, K. & Tadokoro, H. (1978) *Rep. Progr. Polym. Phys. Jpn.* **21**, 417.

Tashiro, K. & Tadokoro, H. (1990) in *Concise Encyclopedia of Polymer Science and Engineering* (J.I. Kroschitz ed.) John Wiley & Sons, New York, p. 187.

Tashiro, K., Hitamatsu, M., Kobayashi, M. & Tadokoro, H. (1986) *Sen'i Gakkaishi* **42**, 659.

Tashiro, K., Nakai, Y., Kobayashi, M. & Tadokoro, H. (1980) *Macromolecules* **13** 137.

Ward, I.M. & Wilding, M.A. (1977) *Polymer* **18**, 327.

Yokouchi, M., Sakakibara, Y., Chatani, Y., Tadokoro, H., Tanaka, T. & Yoda, K. (1976) *Macromolecules* **9**, 266.

Chapter 7

Application of microhardness techniques to the characterization of polymer materials

In the previous chapters the main fields of application of the microhardness technique in polymer physics have been highlighted. The emphasis has been mostly on solving structural problems, looking for relationships between the structures of polymers and their properties (initially mechanical ones) or on studying the factors which determine the microhardness behaviour of various polymeric systems.

This chapter presents selected examples of the application of microhardness measurement for the characterization of polymeric materials after various physical treatments.

These include: the effect of the processing conditions on the mechanical properties of synthetic and natural polymers, the characterization of ion-implanted polymer surfaces, the study of mechanical changes in polymer implants after wear, the influence of coatings on surface properties, weatherability characterization of polymers, etc.

7.1 Effect on microhardness of processing conditions of polymers

The mechanical and physical properties of moulded parts, particularly those made of thermoplastics, do not depend only on the chemical constitution of the material and its properties. The processing conditions also exercise a considerable influence (Wilkinson & Ryan, 1999). Properties such as strength, toughness, hardness may vary to a greater or lesser extent in the same material, or can be selectively varied by choosing a particular processing technique. Those factors that determine the quality of a part are frequently not apparent externally, but are reflected by the internal structure of the part.

The most important structural characteristics of thermoplastics that are dependent on the processing conditions are: (i) molecular orientation, (ii) residual stresses and (iii) crystal structure and degree of crystallinity (of partly crystalline materials).

Possible changes in the molecular structure through a reduction of the chain length or degradation are not discussed here. The factors responsible for this are residence time in the barrel, melt temperature, and an intense shear effect in the runner system during injection.

The results of using microindentation hardness to characterize injection-moulded PE and PET samples prepared using a range injection (processing) temperatures T_p and mould temperatures, T_{mould} will be presented.

7.1.1 Microindentation anisotropy in injection-moulded PE

In the first part of this section we will concentrate on oriented bars of ultra-high-molecular-weight PE prepared by elongational flow injection moulding (EFIM) (Bayer et al., 1989). This is a processing method which yields high-strength materials due to the self-reinforcing effect of the highly oriented fibrils (Bayer et al., 1984). The method uses a mould which induces an elongational flow during the process of injection. The shape and orientation are fixed in the mould by rapidly cooling the injected material. The morphology of EFIM PE is that of shish-kebab fibrils (Ania et al., 1996). It consists predominantly of oriented fibrils (shish cores) and laterally grown, stacked lamellae perpendicular to the injection direction (Fig. 7.1). The influence of temperature on the axial X-ray long periods of melt-processed PE with a shish-kebab structure has been the object of a separate study (Rueda et al., 1997).

The purpose of this application is twofold: (i) to obtain information on the level of orientation across the moulding thickness and the microstructural mechanisms responsible for the observed mechanical anisotropy; (ii) to examine the changes in microstructure, molecular orientation and mechanical properties occurring throughout the range of processing temperatures used.

It will be shown that the level of uniaxial orientation and the variation of local mechanical properties generated by controlling the injection temperature can be conveniently characterized by microhardness measurement in combination with the measurement of optical birefringence Δn and DSC. In Section 2.7 we saw that microhardness is a very useful mechanical property, which can provide direct information about the anisotropy developed within highly oriented polymers.

Figure 7.2 schematically shows the geometry of an injection-moulded polymer bar in which z is the injection direction. Indentations were made on the yz plane. In all cases an indentation anisotropy arises because the microhardness is maximum when the indentation diagonal is parallel to the injection direction (H_z) and minimum when the diagonal is normal to it (H_y). The large H_z value corresponds to

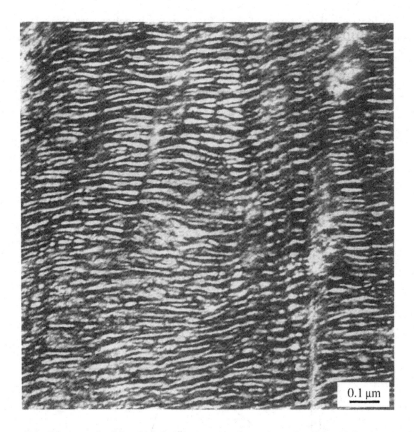

Figure 7.1. Micrograph of the shish-kebab structure from elongational flow injection-moulded high-molecular-weight PE. (From Ania *et al.*, 1996.)

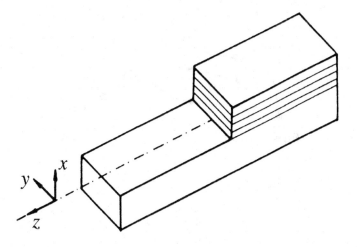

Figure 7.2. Geometry of an injection-moulded polymer bar, where z is the injection direction. Indentations were made on the yz plane. (After Rueda *et al.*, 1989.)

an instant local elastic recovery of shish-fibrils in the injection direction after load release. The low H_y value is ascribed to the plastic deformation of the oriented material (shish-kebabs) (see Section 2.7.2).

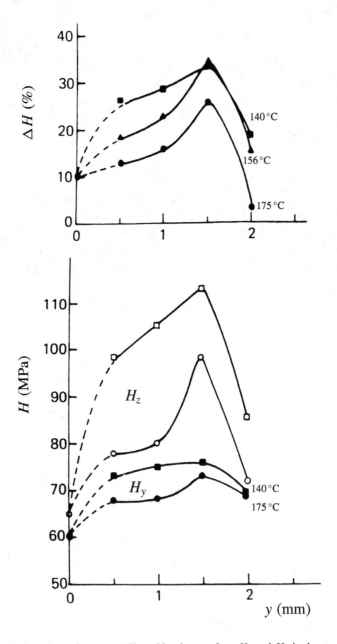

Figure 7.3. Indentation anisotropy ΔH, and hardness values H_z and H_y in the yz plane (Fig. 7.2) at the centre of the PE bar ($x = 2$ mm) measured along the y axis for various processing temperatures T_p. (After Rueda *et al.*, 1989.)

The ΔH, H and Δn variations that occur throughout the moulding thickness are shown in Figs. 7.3 and 7.4 for various processing temperatures.

The results show that the molecular orientation and indentation anisotropy within elongational flow injection moulded PE is low near the surface and in the core of the mouldings, while it is high between these zones. One has to recall at this point that indentation anisotropy ΔH is a consequence of the high orientation of molecular chains coupled with local elastic recovery of the material in the chain direction (see Chapter 2). On the other hand, the optical birefringence $\Delta n = n_1 - n_2$ (n_1 and n_2 being the refractive indices in and perpendicular to the fibre axis, respectively) is the sum of the orientation contributions of crystals and disordered regions. The good correlation obtained between ΔH and Δn (Fig. 7.5) characterizes ΔH as a suitable parameter for measuring the preferred chain axis orientation, which in the present case is a function of injection (processing) temperature T_p and varies across the thickness of the moulded bar. The negative ΔH values observed are only an

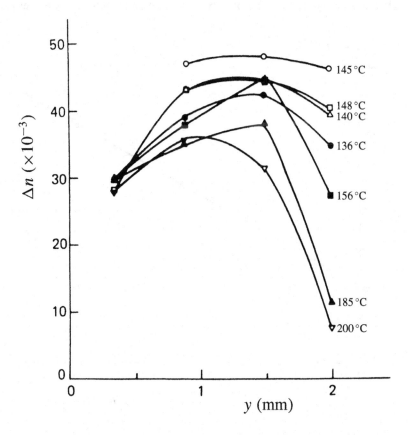

Figure 7.4. Birefringence Δn of 30 μm thick yz cuts (Fig. 7.2) measured at the centre of the PE bar ($x = 2$ mm) along the y axis for selected processing temperatures. (After Rueda *et al.*, 1989.)

artefact which can be attributed to the influence of the highly oriented zone of the bar, at $x = y \sim 1.5$ mm, on the y indentation diagonal, yielding an asymmetric distorted impression and leading to a negative indentation anisotropy value.

Measurements at the surface of the injected bars give values of about $H \sim$ 60 MPa, which are close to the values obtained for linear isotropic PE with a density $\rho = 0.96$ g cm^{-3}. The low orientation at the surface of the mouldings explains the low indentation anisotropy ($\Delta H = 10\%$) shown in Fig. 7.3.

It is to be noted that H at the surface of the bars is nearly independent of T_p. This is because the test entails a penetration depth of only a few micrometres (a soft outer skin) and does not reflect the high-strength fibre structure near the centre. A study of the microhardness of the core, in the z direction reveals a similar type of variation of H with T_p, showing a maximum value around $T_p \sim 145\,^{\circ}$C (Rueda $et\ al.$, 1988). The result that H_y at the core ($\sim$70 MPa) is larger than at the surface ($H_y \sim 60$ MPa) is again due to the presence of the processing skin. The indentation anisotropy can be considered proportional to the number of shish-kebab fibrils contributing to local elastic recovery. The density of fibrils is maximum for mouldings processed at

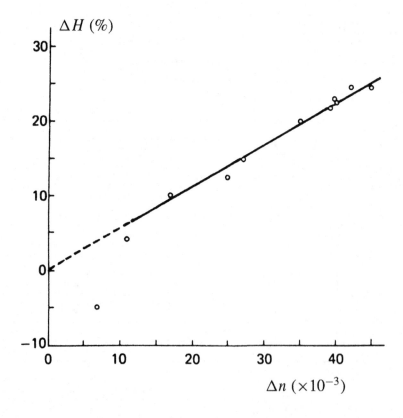

Figure 7.5. Correlation between indentation anisotropy ΔH and birefringence Δn for injection-moulded oriented PE. ΔH and Δn were measured at the centre of the PE bar along the z axis (Fig. 7.2). (After Rueda $et\ al.$, 1989.)

$T_p \sim 150\,^{\circ}\mathrm{C}$. The variation of H_z and ΔH across the thickness of the bars parallels the Δn behaviour, showing a minimum of birefringence at the core, and supports a lower density of shish fibrils in the centre of the mouldings (Figs. 7.3 and 7.4). The decrease of birefringence at the central z axis of the mould is additionally related to thermal contraction effects during cooling and leads to the formation of cracks along the central axis (Bayer $et\ al.$, 1989). The maximum value of $\Delta H = 30\%$ at a depth of about $y = 1.5$ mm is lower than that obtained for solid-state extruded PE (where $\Delta H \sim 40\%$) (Baltá Calleja $et\ al.$, 1980). However, the value of ΔH obtained using elongational flow injection is larger than the value obtained for conventionally injection moulded PE, irrespective of molecular weight (Rueda $et\ al.$, 1981). The coexistence of shish-fibrils and kebab-crystals, contributing to the different values of H_z and H_y, is supported by the presence of two melting peaks in the thermograms (Rueda $et\ al.$, 1989). From the foregoing it is deduced that the formation of shish-crystals in elongational flow injection moulded PE shows a maximum value at about $T_p = 150\,^{\circ}\mathrm{C}$.

In summarizing, it can be concluded that the microhardness of elongational flow injection moulded PE is influenced by a local double mechanical contribution: (a) a plastic deformation of crystal lamellae under the indenter, and (b) an elastic recovery of shish-fibrils parallel to the injection direction after load removal. Further, the Shish-crystals are preferentially formed when high orientation occurs, i.e. at zones near the centre of the mould and at an optimum processing temperature T_p around 145–150 °C. Below this temperature overall orientation decreases due to a wall-sliding mechanism of the rubber-like molten polymer.

7.1.2 The influence of processing parameters on injection-moulded PET

The correlation of microhardness and morphology for injection-moulded PET will be highlighted in this section as a second example of the application of the micro-hardness technique.

The microhardness variations that occur across the moulding thickness have been measured, and the effect of thermal treatment of pellets of the original material on the properties of the mouldings has been analysed. The effect of an annealing treatment on the microhardness of the mouldings has also been examined (Baltá Calleja $et\ al.$, 1993).

The experiments described were performed using pellets of PET with a weight average molecular weight of $\bar{M}_w = 29\,800$ which were used to prepare the injection moulded materials.

The influence of the mould temperature on the H measured at the outer surface of the PET bar is shown in Fig. 7.6. The stepwise increase of H with T_p can be explained because below and near T_g, H values between 120–130 MPa have been ascribed to the incipient spherulite structure of the polymer (see Section 4.2.2,

Fig. 4.8(a)). On the other hand, for mould temperatures well above T_g, the microhardness reaches values of about 200 MPa. These high H values correspond to structures in which the material is fully covered by the crystallizing units (spherulites) (Fig. 4.8(b)). Figure 7.7 illustrates the influence of the mould

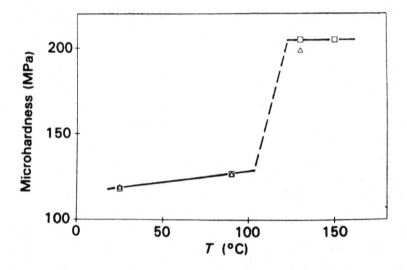

Figure 7.6. Microhardness of injection moulded PET, measured at the outer surface, as a function of mould temperature. The injection was performed using: conventional samples as received ($\triangle$); dried samples ($\square$). (From Baltá Calleja *et al.*, 1993.)

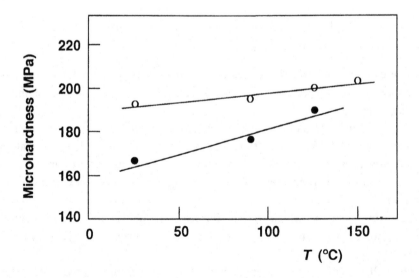

Figure 7.7. Microhardness measured at the inner yz surface (see Fig. 7.2) on PET bars processed at various mould temperatures: samples as received ($\blacksquare$) and samples dried at 150 °C under vacuum ($\square$). (From Baltá Calleja *et al.*, 1993.)

temperature on the values of H measured on the inner surface. One sees that larger hardnesses are obtained if the pellets are dried before injection moulding. Figure 7.8 shows the H values observed on the inner yz surface across the thickness of the moulding (the y direction) for various PET samples moulded at different temperatures. The gradual improvement of microhardness with increasing mould temperature T_{mould} has been ascribed to an increasing volume fraction of spherulitic material. Because of the slow crystallization rate for PET, for the samples moulded at $T_{mould} = 25\,^\circ$C the crystallization within the bar is not as complete as it is in the case of $T_{mould} = 120\,^\circ$C where cooling proceeds more slowly. Therefore, in the case of $T_{mould} = 25\,^\circ$C the samples are only partially filled with spherulites and their H values are less than 200 MPa.

The H profile in the y direction supports the presence of an outer softer amorphous layer ($H = 120$ MPa) in the samples with T_{mould} below 120 $^\circ$C (Fig. 7.8). However, on increasing T_{mould} above 120 $^\circ$C, the amorphous layer crystallizes and hardens. It should be noted that for the samples prepared at $T_{mould} \geq 120\,^\circ$C, H shows a distinct maximum value at both surfaces, suggesting an enhancing nucleation effect of the metallic walls of the mould on the crystallizing material.

In general the use of low mould temperatures leads to mouldings with low H values in which primary crystallization is not completed (Baltá Calleja et al., 1993). One way to improve the properties of the injection-moulded material is by means of a subsequent annealing treatment at temperatures, T_a, well above $T_g = 70\,^\circ$C.

It is known (Friedrich & Fakirov, 1985) that injection moulding of PET which has not been carefully dried results in a serious degradation of its mechanical properties profile due to hydrolitic degradation of the molecular weight, leading to an increase

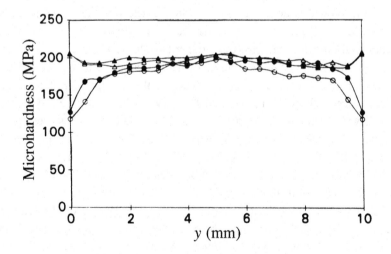

Figure 7.8. Microhardness profiles along the y direction (Fig. 7.2) obtained from measurements at the inner yz surface of samples for T_{mould}: (O) 25 $^\circ$C; (●) 90 $^\circ$C; (△) 130 $^\circ$C; (▲) 150 $^\circ$C. (From Baltá Calleja et al., 1993.)

in the crystallinity after processing. This explains the differences in H for dried and non-dried PET material obtained by Baltá Calleja et al. (1993) (see Fig. 7.7).

7.2   Characterization of natural polymers

7.2.1   Starch

In recent years starch, the polysaccharide of cereals, legumes and tubers, has acquired relevance as a biodegradable polymer and is becoming increasingly important as an industrial material (Fritz & Aichholzer, 1995). Starch is a thermoplastic polymer and it can therefore be extruded or injection moulded (Baltá Calleja et al., 1999). It can also be processed by application of pressure and heat. Starch has been used successfully as a matrix in composites of natural fibres (flax, jute, etc.). The use of starch in these composites could be of value in applications such as automobile interiors. An advantage of this biopolymer is that its preparation as well as its destruction do not act negatively upon the environment. A further advantage of starch is its low price as compared with conventional synthetic thermoplastics (PE, PP).

All starches are biosynthesized as semicrystalline microgranules containing densely packed polysaccharides (amylose and amylopectin) and relatively little water (Poutanen & Forssell, 1996). The microgranules which have dimensions ranging typically from 1 to 100 μm, are often compared to polymer spherulites with a radial orientation of the starch polymer chains. It is commonly accepted that amylopectin, the high-molecular-weight (10^8 g mol^{-1}), highly branched constituent of starch microgranules, is chiefly responsible for the crystalline structure of native starch. Studies have shown, in addition, that the microstructure of native starch grains is characterized by the alternation of semicrystalline amylopectin, shell-shaped zones and amorphous, amylose regions (Cameron & Donald, 1993)

Interpretation of the WAXS patterns of native starch is often difficult because of the low crystallinity, small size, defects and the multiple orientations of the amylopectin crystallites (Waigh et al., 1997). Two main types of X-ray scattering patterns have been commonly observed (A and B). Potato starch has been shown to crystallize in a hexagonal unit cell in which the amylopectin molecules twist in a double helix (the B structure) (Lin Jana & Shen, 1993). Between adjacent helices a channel is formed in which 36 water molecules can be located within the crystal unit cell. By means of heat treatment this structure can be transformed into a more compact monoclinic unit cell (the A structure) (Shogren, 1992). Amylose (the linear and minor component of starch) can be crystallized from solution in the A and B structures (Buleón et al., 1984), yielding X-ray diffraction patterns similar to those of amylopectin but with higher orientation.

Starch is still at the initial stage as far as its applicability as a polymeric material is concerned. Preliminary experiments have been reported on the development of high-strength and high-modulus PE–starch composite films and the degree of biodegradability has been examined (Nakashima & Matsuo, 1996). Studies (Baltá Calleja *et al.*, 1999) have revealed the influence of processing methods on the structure and properties of potato starch.

It is well known that every starch processing method is facilitated by the presence of water. Water depresses the melting point of amylopectin crystallites and lowers starch melt viscosity. Native potato starch contains about 18% water when stored at ambient atmosphere (room humidity about 55%). If one does not add more than about 10% water the melt-processed starch will solidify when cooling down. If more water is added to the starch, its glass transition falls below room temperature (Zeleznak & Moseney, 1987). The moulding then exhibits a rubbery state and solidifies later when the excess water evaporates.

In the above studies compression and injection moulding were carried out with native potato starch dried at room temperature and then mixed with 7% water. After processing, the samples were subjected to ambient atmosphere and investigated after a storage time of at least 4 weeks.

Compression moulding of the starch granules leads to sintered relatively brittle materials. Here the amylopectin crystals of the native powder largely remain

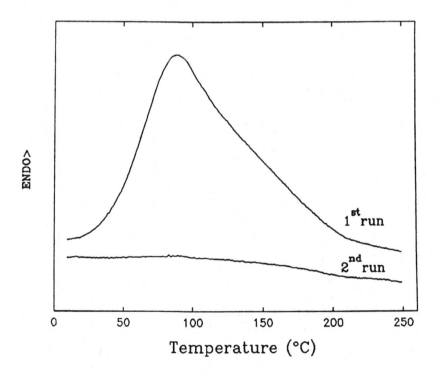

Figure 7.9. First and second temperature DSC run for an injection moulded starch sample. (From Baltá Calleja *et al.*, 1999.)

preserved. Preparation of dry films from aqueous gels results in a disintegration of the structure of the native starch granules and in the formation of a new semi-crystalline structure comprising crystallized amylose molecules. Injection moulding of native starch is found to be a processing method which gives rise to amorphous materials with superior mechanical properties.

As the X-ray diffraction patterns show no orientation, one may speculate that the higher mechanical strengths obtained with elongational flow moulding are due to stretching of the amylose network, which finally relaxes in the mould. In addition, mechanical activation during processing favours gelatinization. A typical thermogram of an injection moulded starch is shown in Fig. 7.9. This shows a very broad maximum at about $100\,^{\circ}\mathrm{C}$ which is related to the evaporation of water. However the high-temperature tail suggests the presence of water molecules that are strongly bound to the starch molecules.

Figure 7.10 shows the microhardnesses of several injection moulded starch samples processed at different temperatures with different initial water content. There is an increase in H with the injection moulding temperature in the range 80–110 °C, from 120 MPa to 140 MPa (Baltá Calleja *et al.*, 1999). These microhardness values are notably higher than those found for conventional injection moulded thermoplastic polymers like PE (50–60 MPa) (Rueda *et al.*, 1989; Baltá Calleja *et al.*, 1995).

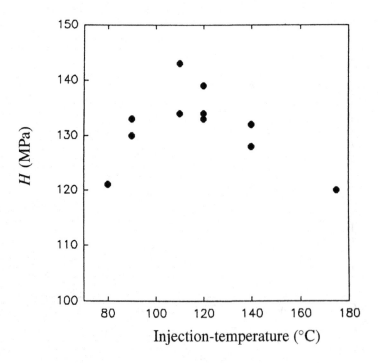

Figure 7.10. Microhardness H of starch samples as a function of injection temperature. (From Baltá Calleja *et al.*, 1999.)

This initial increase in microhardness with temperature might be connected with the gradual disappearance of defects and grain boundaries. For processing temperatures T_p higher than $110\,^{\circ}\text{C}$ a decrease in microhardness is observed. The H variation with temperature is linearly related to the corresponding changes observed in the density of the samples (Secall, 1996). The decrease in H observed for $T_p > 110\,^{\circ}\text{C}$ is probably related to some degradation of the material at higher temperatures (Harper, 1992).

It has been also observed that the microhardness of the mouldings rises significantly after removing the water from the sample. Water is the usual plasticizer in starch processing and mechanical properties are greatly influenced by the water content (Poutanen & Forssell, 1996). Figure 7.11 illustrates the variation in the microhardness after storing the dry sample at ambient atmosphere. The dry sample has a microhardness of about 210 MPa which is comparable to that of some metals (Al, Cu, Ag, etc.) (Paplham et al., 1995). However, the microhardness value rapidly decreases with storage time and after 1 day it has returned to that of an air-stored sample.

In conclusion, only through the thermomechanical processing of starch by means of injection moulding can the initial semicrystalline microgranule structure be destroyed and a homogeneous amorphous high-performance material be obtained. This material displays a better performance when processed by using an elongational

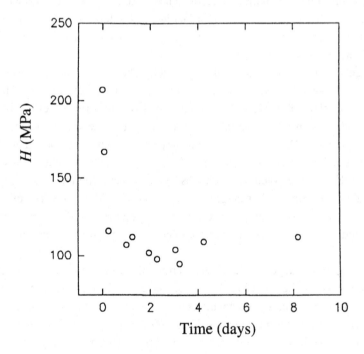

Figure 7.11. Microhardness H of dry starch samples in ambient atmosphere as a function of storage time. (From Baltá Calleja et al., 1999.)

flow mould. Microhardness measurements of the injection mouldings evidence mechanical properties which are close to those typical for some soft metals.

7.2.2   Gelatin

Gelatin is a protein which has been used as a glue from ancient times. In a study on the melting behaviour of gelatin film, it was found that this rather common biopolymer is characterized by very high surface microhardness (Fakirov et al., 1999). It turns out that in this respect it surpasses all commercial synthetic polymers, exhibiting microhardnesses varying typically between 100 and 250 MPa (Baltá Calleja & Fakirov, 1997). Room conditioned gelatin, i.e. gelatin containing about 15 wt% water, which has strong plasticizing (softening) effect as concluded from the drop of the glass transition temperature T_g (from 217 °C for dry samples to 60 °C) (Rose, 1987), shows a relatively high microhardness – above 200 MPa. With increasing temperature at which the H measurements are carried out the microhardness gradually increases, reaching 390 MPa in the range 130–150 °C. It is assumed that this doubling in the microhardness is correlated both to the crystallization process and to the dry state of the sample – at 130–150 °C gelatin is completely dried, as found by DSC and gravimetric measurements (Fakirov et al., 1997).

This strong effect of temperature on microhardness motivated further interest in the temperature dependence of microhardness of gelatin at higher temperatures, covering the ranges of glass transition and melting of dry gelatin (217 °C and 235 °C, respectively) (Vassileva et al., 1998).

The gelatin films were prepared from food grade powder, soaked overnight at 5 °C, then dispersed in water at 50 °C and cast in aluminium dishes. The room conditioned gelatin film (15–17 wt% water content) was further dried for 5 h at 140 °C under vacuum and thereafter moistened to 11 wt% water content in order to depress the T_g. The sample was then subjected to annealing for 5 h at 90 °C in a vessel containing water to produce low-melting crystallites, followed by drying for 5 h at 80 °C under vacuum in the absence of water. The film thus treated was repeatedly tested on the hot stage starting at 100 °C. Measurements of H were performed on the same sample at different temperatures in the interval between 100 °C and T_g with the sample being cooled down to room temperature between each cycle.

The temperature dependence of the microhardness for the gelatin film treated as described above is shown in Fig. 7.12. For the sake of comparison, the H (around 200 MPa) of room conditioned gelatin is also given (Fig. 7.12(a), the open circle).

One can see that the first heating from 100 to 191 °C (Fig. 7.12(a)) results in a microhardness increase from 330 to 450 MPa. During heating up to 205 °C (Fig. 7.12(b)), a further 23% increase in microhardness is observed, as compared to the maximum value reached in the first measurements (Fig. 7.12(a)). The third series of measurements starts at a microhardness of about 530 MPa and reaches its highest value of 654 MPa at 196 °C after which it drops up to 510 MPa (Fig. 7.12(*c*)) thus confirming the tendency of microhardness to decrease at temperatures above 200 °C. The observed decrease in the microhardness around 200 °C, i.e. close to T_g of dry gelatin, is presumably related to the softening of the sample. The last curve (Fig. 7.12(*d*)) shows an almost constant microhardness value (between 630 and 565 MPa) for the temperature range of 100–200 °C followed by a large drop to 390 MPa over the next 50 °C (200–250 °C). This is most probably due to softening effects in the material above its T_g.

What could be the reason for the observed extremely high values of microhardness in the gelatin films measured at relatively high temperatures, approaching T_g and T_m of dry gelatin (Fig. 7.12)? In order to answer this question, one has to take into account the chemical peculiarities of this polymer. Being a polypeptide, rather than a chemically similar polyamides, the gelatin is characterized by the presence of a large amount of free (side chain) carboxyl, hydroxyl and amino groups arising from the diaminomonocarboxylic and monoaminodicarboxylic acids.

Condensation polymers are well known to undergo additional condensation (Flory, 1953; Fakirov, 1990, 1999) involving the end groups, when appropriate con-

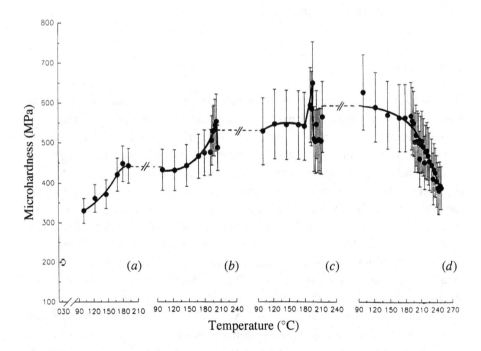

Figure 7.12. Dependence of microhardness of gelatin on the temperature at which it is measured. The *H* measurements are carried out on the same sample which was stored in a desiccator between the temperature cycles. The open circle in (*a*) corresponds to the room conditioned (not-dried) gelatin film. (From Vassileva *et al.*, 1998.)

ditions exist (temperature, catalysts and vacuum are the main factors accelerating the process). Similar interactions between reactive side chain groups can be expected in the case of proteins as frequently reported (Yannas & Tobolsky, 1967; Ward, 1977). For instance, the insolubility of gelatin after sufficiently prolonged (days!) evacuation at temperatures between 65 and 105 °C is explained by the formation of a three-dimensional network resulting from interchain cross-linking (Yannas & Tobolsky, 1967). This conclusion is supported by the fact that such an insolubility is not observed when chemically modified (by acetylation of the amino groups or by esterification of the carboxyl groups) gelatin is subjected to the same treatment (Bello & Riese-Bello, 1958).

One can assume that as a result of formation of chemical links between chain segments, regardless of whether these are intra- or intermolecular, a denser chain packing will be achieved, compared to that of chemically non-bonded chains or chain segments. Such a densification will lead to increased microhardness since it is known that these two properties are very closely related (Baltá Calleja, 1985; Baltá Calleja et al., 1987).

In conclusion, one has to emphasize that thermally treated gelatin films exhibit surface microhardnesses that are much larger than those of the common synthetic polymers. This is most probably due to the occurrence of chemical cross-linking in gelatin as a result of thermal treatment.

7.2.3   Keratin

Keratin, the basic substance of hair, is another protein. Individualizing human hair from a very small amount of material is desirable for forensic medicine (Paar, 1990a). Hair microhardness, as well as a description of hair thickness, colour and morphology, can be very helpful in distinguishing hair from different people. Also the hair of those who are exposed to a particularly hostile environment, either physical (temperature, humidity) or chemical (vapours, gases, toxic heavy metals), can undergo microhardness changes. The cross-section of a human hair cut using a usual histological method yields a surprisingly smooth surface for the indentations. Scalp hair (one single piece of 1 cm length is sufficient) is embedded in resin and cut with a microtome. The cuts (five per person) are mounted on a microscope slide without any further preparation. Their surface is quite smooth except for the foamy medulla. Onto these cross-sections Vickers indentations (five per hair) with a load of 2 g are set (Fig. 7.13). Seven diagonal readings are made of each imprint. Statistical evaluations are performed directly on the diagonal readings. The following questions arise: do different indentations on the same hair lead to the same microhardness value? Is the statistical scatter in the microhardness of one individual sufficiently small, and is the average microhardness significantly different for different individuals? The answers can be found in Figs. 7.14 and 7.15.

Despite the low refractive index of human hair the contrast is sufficient for microscopic observation. In its cortex a human hair has a microhardness in the region of 20 MPa. The reproducibility of the readings for a single indentation is very high and can be seen in Fig. 7.14. The relative standard deviation of all indentations from one person is also very low. The microhardness of hair correlates

Figure 7.13. Human hair cross-section; Vickers imprint with a 2 g load. (From Paar, 1990.)

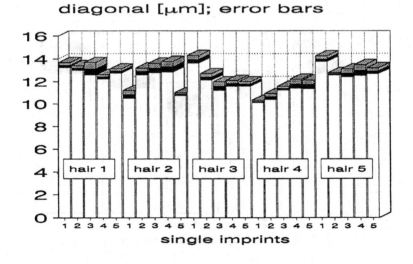

Figure 7.14. Indentation measurements on different hairs of the same person. Each indentation imprint corresponds to seven measurements. (From Paar, 1990.)

well with individuals especially in connection with statistical evaluation methods (Fig. 7.15).

7.3   Mechanical changes at polymer surfaces

7.3.1   Wearing of polymer implants

Ultra-high-molecular-weight PE has been used with success for over 20 years to provide a low-friction articulating surface in total joint replacements (Morcher, 1984). In later developments, hydroxyapatite–PE composites have been used in orbital surgery (Downes *et al.*, 1991). Kalay *et al.* (1999), through the application of shear controlled orientation in injection moulding (Kalay *et al.*, 1997) to high-density PE, achieved a fourfold increase in E modulus (7.2 GPa) and fivefold increase in the ultimate tensile strength (155 MPa) with respect to conventional injection moulding. This material has very good prospects for orthopaedic applications.

Although replacement surgery is increasingly used for hip joints, these still suffer from limited lifetimes in the body due, at least in part, to the wear of the implanted polymer.

In the context of widespread research aimed at understanding the wear processes in order to improve the materials and so prolong the effectiveness of the implant, researchers have examined a number of PE hip cups worn on a cobalt chrome femoral head before and after implantation in the body (Olley *et al.*, 1999). In early investigations the use of permanganic etching prior to light and electron microscopy was shown to allow sensitive discrimination between the quality of

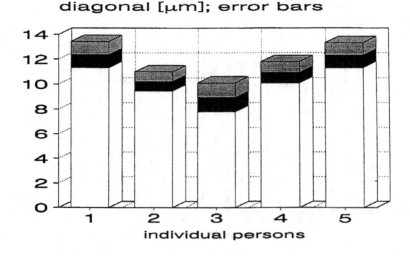

Figure 7.15. Indentation diagonals from hairs of different people, except nos 1 and 5, which are identical and come from the same person. (From Paar, 1990.)

different moulded materials prepared for possible use in the manufacture of hip cups.

More recently it has been possible to examine used hip cups which have been removed from the body. Electron microscopy of hip cup sections after permanganic etching reveals a layer of polymer, at least 10 µm thick extending inwards from the wear surface, which showed incipient cracking and appeared to have been embrittled by exposure to the body environment (Jordan *et al.*, 2000).

Microhardness tests upon the same worn hip cups have provided direct supporting evidence for the changed nature of PE adjacent to the wear surface (Flores *et al.*, 2000). An increased microhardness of the used cups, at their wear surfaces, in comparison to the control was apparent. Attention turned, accordingly, to the lateral surfaces of the cups (Fig. 7.16) which did possess the smooth planar condition required for optimum microhardness measurements. Figure 7.16 shows microhardness measured as a function of the radial distance, h, from the indent to the edge of the concave surface. The microhardness of the control sample ($H \approx 57$ MPa) does not vary with h. however, H for the hip cups, after implantation and removal from

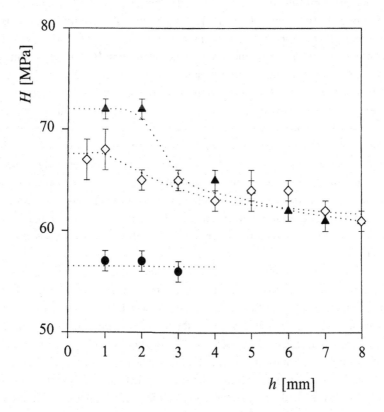

Figure 7.16. Variation of microhardness as a function of the distance to the concave surface of a hip cup: before implantation (●); after 36 months implant (◇); after 73 months implant (▲). (From Flores *et al.*, 2000.)

the body, shows a stepwise increase as the wear surface is approached. An initial H plateau, up to $h \sim 1$–2 mm, has been brought about by *in situ* conditions at the hip joint. Also notable is the fact that the microhardness in both worn samples at $h > 4$ mm is still higher than that of the non-implanted material.

The above results support previous electron microscope observations showing incipient cracking and brittleness adjacent to the wear surfaces of PE hip cups after exposure to the body environment by showing suitably altered mechanical properties. A possible mechanism to account for the observed increase of micro-hardness in the implanted materials (at $h > 4$ mm) is (oxidative) degradation with incorporation of inorganic atoms in the amorphous (interlamellar) component of PE which, according to the measured crystallinity values comprises about 50% of the material. It has been shown that chemical treatment of LDPE with OsO_4 produces a drastic increase in microhardness which has been explained in terms of a reduction in the molecular mobility of the amorphous region (Baltá Calleja *et al.*, 1997). A similar mechanism, leading to a slight hardening of the amorphous phase, could well take place in the materials investigated here.

For quantitative evaluation of the microhardening effect within the amorphous regions, using eq. (4.3) with $H = 57$ MPa and degree of crystallinity $w_c = 0.517$, and assuming $H_a \sim 0$ for the control sample, a value for H_c of 110 MPa is obtained. In accordance with Fig. 7.16 the crystal microhardness H_c and crystallinity w_c are assumed to remain constant after implantation when $h > 4$ mm. Using these values together with eq. (4.4) and taking $H \sim 64$ MPa when $h > 4$ mm for the used samples, one derives a value of $H_a = 15$ MPa. Hence, the microharden-ing effect observed on the implanted samples at $h > 4$ mm, can be correlated with the microhardness increase of the amorphous regions, from $H_a \sim 0$ in the non-implanted material to $H_a \sim 15$ MPa after exposure in the body. The step-wise increase in microhardness of the samples after wear, as the wear surface is approached ($h = 0$) (Fig. 7.16), is in agreement with the increases in crystallinity observed with respect to the control sample. The sample with the longest implant duration, shows the highest degree of crystallinity and, consequently, has the highest microhardness.

These results demonstrate that the embrittlement of the PE implants accompanies a microhardening of a surface layer and an increase in crystallinity. The two pieces of evidence are complementary and imply a reduction in the crack-blunting ability of the material, i.e. a diminution of the number of interlamellar tie molecules which connect adjacent lamellar stacks. In consequence the elastic properties of the material diminish and cause the material to microharden during wear. The increase in microhardness at the wear surface is partly because the amorphous component decreases in quantity and partly because its chemical nature changes as it undergoes simultaneous microhardening and loss of elasticity.

7.3.2 Characterization of irradiated and ion-implanted polymer surfaces

During the past few years the microhardness technique has frequently been applied to the characterization of super-hard-surfaced polymers obtained by ion implantation and to plasma-deposited hard amorphous carbon films (Baltá Calleja & Fakirov, 1997). These products represent an entirely new class of materials that are lightweight and have the flexibility of polymers combined with a surface microhardness and wear resistance greater than those of metallic alloys (Lee *et al.*, 1996).

The surfaces of *PE*, *PC* and poly(ether imide) (PEI) have been modified by implantation of metallic ions from a metal-vapour arc source (Rao *et al.*, 1994). Nanoindentation hardness and surface resistivity changes showed that Ti and Si implantation caused a similar microhardness increase for PC and PEI, while Cr implantation resulted in the largest microhardness improvement for PE and PC. This was attributed to there being more chain scission caused by nuclear collisions of the heavy Pt ions than cross-linking caused by electronic excitation. In the case of PE, Ti and Cr yielded similar microhardness increases although the measured microhardnesses were smaller than those for PC and PEI. The implantations also decreased the electrical resistivity of the polymers. Ti, Si and Cr implantations exhibited trends for a decrease in resistivity similar to those for the increase in microhardness suggesting a similar underlying cause, namely cross-linking. It appears that irradiation-induced cross-links provide paths for increased mobility of carriers contributing towards decreased resistivity. Pt implantation yielded the largest increase in conductivity, partly as a result of the high dose used and partly because of the deposition of microdroplets of Pt which form at the metal ion source causing Pt to accumulate at the polymer surface leading to metallic conduction. This study represented the first use of a mixed energy spectrum of ion implantation to modify surface properties of polymers (Rao *et al.*, 1994).

Another set of polyolefins including PP, PS and poly(ether sulphone) (PES), were implanted with 200 keV B ions at three different doses by Rao *et al.* (1992). PS was also implanted with 200 keV B ions. The nanoindentation technique and reciprocating sliding wear tests were used to characterize the properties of the implanted polymers. The results show that microhardness increases with increasing dose as well as with increasing energy. The percentage increase in microhardness reduced with increasing complexity of the side groups although the absolute microhardnesses were higher. With increasing complexity of the main backbone of the polymer chain, the percentage increase in microhardness, as well as the microhardness value, was lower. These effects are attributed to the relative contributions of cross-linking and chain scission which are viewed as competing processes. No specific trends were observed for relating friction coefficient values and improved wear resistance. The results indicate that wear is a complex phenomenon which has a sensitive dependence on the microhardness and elastic properties of the mating surfaces (Rao *et al.*, 1992).

To explore the behaviour of ion-implanted polymers microfriction studies were conducted on 1 MeV Ar$^+$ implanted PEEK, and PS implanted to fluences of 5, 10 and 50 $\times$ 10^3 ions m^{-2} (Rao *et al.*, 1995). The results were compared with macrofriction values obtained using standard pin-on-disc-type tests. The polymers were also characterized for surface mechanical properties using the nanoindentation technique. The most striking aspect of the microfriction tests on the ion-implanted polymers is a marked 'stick–slip' behaviour, which was not observed for the non-irradiated polymers.

This friction behaviour has been related to the three-dimensionally cross-linked structure of the ion-irradiated polymer material, since, with increasing fluence, the polymer surface becomes harder and less elastic due to a greater extent of cross-linking. The stick–slip behaviour is thus caused by the adhesion of the two surfaces, and the periodic elastic extension and sudden release of the cross-linked structure of the implanted layer. The microfriction results have been correlated with nanoindentation hardness measurements, which are an indirect measure of the extent of cross-linking (Rao *et al.*, 1995).

The microhardness variations of polyisoquinoline, poly(2-vinylpyridine) and polyacrylonitrile, induced by irradiation with increasing fluences of Au, Ne, C, Li, He or H ions, have also been measured by Pivin (1994), by means of the nanoindentation test. The resulting stiffening of the structure is more important than that arising from a pyrolysis treatment, because of a transformation in the three-dimensional network with a higher coordination of each species. The dominant factor in microhardness enhancement is the amount of energy dissipated in ionization, while the breaking of chains by nuclear collisions has no significant effect. The efficiency of ionization depends on the spatial density (determined by the mass and energy of ions), indicating that the radicals formed at each impact tend to recombine when they cannot interact.

The hardening and embrittlement of polyimides by ion implantation has been also studied (Pivin, 1994). Nanoindentation tests performed using a sharp diamond pyramid of apical angle 35° provided very quantitative depth profiles of microhardness in polyimides implanted with C, N, O, Ne or Si ions. In all cases the microhardness increased steeply when the amount of deposited energy reached the order of 20 eV atom^{-1}. For energies of 200 eV atom^{-1} the polymer is transformed into an amorphous hydrocarbon and the microhardening factor saturates at a value of 13–20. However, the carbonized layer is poorly adherent, as is evidenced by reproducible discontinuities in the depth *vs* load curves, when the indenter tip reached the interface.

7.4   Weatherability characterization of polymer materials

7.4.1   Very-near-surface microhardening of weathered window-grade PVC

Creep and microhardness measurements are used today to study the environmental stress crack (ESC) resistance of PMMA and unplasticized PVC (UPVC). Monotonic creep is shown to discriminate, to a high resolution and in the short term, the ESC resistance of polymer/fluid pairs, including those polymer/fluid pairs which exhibit mild/weak interactions. Microhardness is shown to offer a cost-effective method of mass screening polymer/fluid pairs for compatibility. The nanoindentation technique has been used by Turnbull & White (1996) to provide unique information on the microhardness and modulus of artificially weathered window-grade UPVC at the submicrometre level. Indentations were made to depths ranging from about 140 nm to 5.5 μm. The sensitivity of the microhardness to loading rate and to hold-time at maximum load has been investigated. Hold-time was the key variable and an optimum period of 200 s was recommended. The microhardness of the artificially weathered window-grade UPVC varied with exposure period and with depth. At long exposure times, the microhardness of the top 147 nm was very much greater than that of the underlying material, which had softened with exposure. This very-near-surface microhardening, which is attributed to cross-linking, cannot be detected by conventional techniques. After extended exposure, evidence of the recovery of properties was observed, consistent with the mechanism of surface layer removal by erosion. The modulus is less affected by weathering than the microhardness with a modest decrease being observed only at very prolonged exposures.

7.4.2   Damage characterization of carbon-reinforced composites

The effect of immersion in seawater on the mechanical properties of graphite/epoxy composite materials has been studied by Grant & Bradley (1995). The transverse tensile strength was found to be reduced by 17% in one of the three systems studied with essentially no change in the other two. Direct observation of fracture using an SEM as well as microindentation measurements of the interfacial shear strength have been used to explain these observations. The 17% decrease in transverse tensile strength has been associated with degradation of the interfacial strength, resulting in a change in the fracture mechanism from primarily matrix cracking to interfacial failure.

Kucner & McManus (1994) investigated the response of advanced graphite-fibre-reinforced epoxy laminated composites, of three different thicknesses, to flame from a calibrated propane burner for times of 3–60 s, by visual observation, light microscopy and planimetric measurements of the area of matrix loss and mass loss on the exposed face of the laminates. The damaged specimens were also examined using a microhardness tester and an SEM. The microhardness tests quantified the degradation of matrix. The microhardness at various points in the cross-section of the burnt specimen, and detailed SEM photographs of the damage indicated the depth of damage and the extent to which the specimens self-insulate. Thin laminates damage faster than thick laminates. Laminate microhardness begins to decrease at a point approximately 5 mm from the location at which the ply was completely destroyed by matrix loss. This decrease in microhardness is not only reflected in cracks, delaminations and other visible types of damage, but also in degradation of the properties of the matrix. Thus, microhardness tests can reveal changes in the laminate which visual inspection cannot. These results help to quantify the advantages and disadvantages of graphite-fibre-reinforced epoxy structures involved in aircraft fires (Kucner & McManus, 1994).

7.4.3   Weathering of i-PP pipes

There have also been weathering studies on i-PP pipe material (Paar, 1998). Standardized burst tests reveal ageing effects on an accelerate time scale. Ageing shows up mainly in enhanced crystallinity of the polymer and is associated with an increase in microhardness.

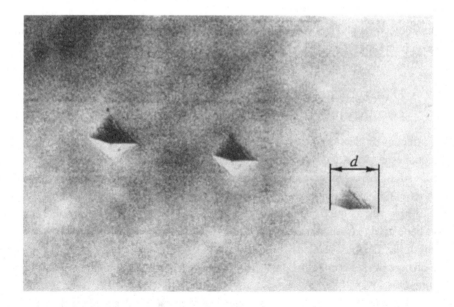

Figure 7.17. SEM micrograph of microindentations on unpigmented i-PP; ageing without internal pressure for 2500 h at 95 °C. (From Paar, 1988.)

The samples have been cut out of pigmented and unpigmented *i*-PP both pipes before and after the burst tests. Their surfaces were covered with a thin layer of Au–Pd in order to ensure electric conductivity for the SEM investigation. Loads of around 7 mN were applied and average microhardnesses of 50 MPa have been obtained along the cross-section of the pipe wall. The standard deviation was 9%. From the *H* vs w_c plot in Fig. 4.7 and the above microhardness value, crystallinities lower than 50% can be estimated for the pipe sections. Fig. 7.17 shows the indentations on unpigmented *i*-PP after ageing at 95 °C for 2500 h. A collection of data obtained from similar tests but on pipes exposed to different internal pressures and ageing temperatures is shown in Fig. 7.18. The data show an increase in microhardness after accelerated ageing. Ageing under internal pressure causes an additional microhardness gain as a result of plastic flow.

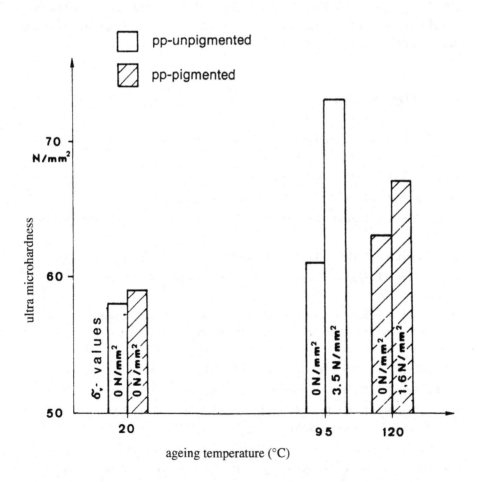

Figure 7.18. Ultra-microhardness of *i*-PP pipe material exposed to different internal pressures (σ); ageing time 2500 h at various temperatures. (After Paar, 1988.)

7.5   Conclusions

In conclusion, the selected examples in this chapter emphasize areas of polymer research that offer new possibilities for applications of the microindentation method to measurements of the mechanical and other properties of polymer surfaces. As such the microhardness technique both extends and complements other modern physical methods used to study and characterize polymer materials. It is specific in suggesting further microhardness morphology correlations of flexible and rigid crystallizable polymers, microfibrillar materials and non-crystallizable glasses. Researchers interested in surface properties will recognize future opportunities in both the characterization of ion-implanted polymer surfaces and coatings and the weathering characterization of polymeric materials. Of particular interest is the applicability of the technique to new high-tech materials characterized by an extremely high surface microhardness. In addition, it is expected that nanoindentation techniques will offer new ways of studying the elastic and plastic properties of the near-surface region of polymers.

7.6   References

Ania, F., Baltá Calleja, F.K. Bayer, R.K., Tshmel, A., Naumann, I. & Michler, G.H. (1996) *J. Mater. Sci.* **31**, 4199.

Baltá Calleja, F.J. (1985) *Adv. Polym. Sci.* **66**, 117.

Baltá Calleja, F.J. & Fakirov, S. (1997) *Trends Polym. Sci.* **5**, 246.

Baltá Calleja, F.J., Baranowska, J., Rueda, D.R. & Bayer, R.K. (1993) *J. Mater. Sci.* **28**, 6074.

Baltá Calleja, F.J., Giri, L., Michler, G.H. & Naumann, I. (1997) *Polymer* **38**, 5769.

Baltá Calleja, F.J., Giri, L., Ward, I.M. & Cansfield, D.L.M. (1995) *J. Mater. Sci.* **30**, 1139.

Baltá Calleja, F.J., Martínez-Salazar, J. & Rueda, D.R. (1987) *Encyclopedia of Polymer Science and Engineering* (Mark, H.F., Bikales, N.M., Overberger, C.G. & Menges, G. eds.) second edition, John Wiley & Sons, New York, p. 614.

Baltá Calleja, F.J., Ohm, O. & Bayer, R.K. (1994) *Polymer* **35**, 4775.

Baltá Calleja, F.J., Rueda, D.R., Porter, R.S. & Mead, W.I. (1980) *J. Mater. Sci.* **15**, 765.

Baltá Calleja, F.J., Rueda, D.R., Secall, T., Bayer, R.K. & Schlimmer, M. (1999) *J. Macromol. Sci. Phys.* **B38**, 461.

Bayer, R.K., Eliah, A.E. & Seferis, J.C. (1984) *J. Polym. Eng. Rev.* **4**, 201.

Bayer, R.K., Zachmann, H.G., Baltá Calleja, F.J. & Umbach, H. (1989) *Polym. Eng. Sci.* **29**, 186.

Bello, J. & Riese-Bello, H. (1958) *Sci. Indust. Photograph* **29**, 361.

Buleón, A., Duprat, F., Body, F.P. & Chanzy, H. (1984) *Carbohydr. Polym.* **4**, 161.

Cameron, R.E. & Donald, A.M. (1993) *J. Polym. Sci., Polym. Phys. Ed.* **31**, 1197.

Downes, R.N., Vardy, S., Tanner, J.E. & Bonfield, W. (1991) *Bioceramics* **4**, 23.

Fakirov, S. (1990) in *Solid State Behaviour of Linear Polyesters and Polyamides* (Schultz, J.M. & Fakirov, S., eds.) Prentice Hall, Englewood Cliffs, New Jersey, p. 1.

Fakirov, S. (ed.) (1999) *Transreactions in Condensation Polymers*, Wiley-VCH, Weinheim.

Fakirov, S., Cagiao, M.E., Baltá Calleja, F.J., Sapundjieva, D. & Vassileva, E. (1999) *Int. J. Polym. Mater.* **43**, 195.

Fakirov, S., Sarac, Z., Anbar, T., Boz, B., Bahar, I., Evstatiev, M., Apostolov, A.A., Mark, J.E. & Kloczkowski, A. (1997) *Colloid Polym. Sci.* **275**, 307.

Flores, A., Baltá Calleja, F.J., Jordan, N.D., Bassett, D.C., Olley, R.H. & Smith, N. (2000) *Polymer* **41**, 7635.

Flory, P.J. (1953) *Principles of Polymer Chemsitry*, Cornell University Press, Ithaca.

Friedrich, K. & Fakirov, S. (1985) *J. Mater. Sci.* **20**, 2807.

Fritz, H.G. & Aichholzer, W. (1995) *Starch* **47**, 475.

Grant, T.S. & Bradley, W.L. (1995) *J. Compos. Mater.* **29**, 852.

Harper, J.M. (1992) in *Developments in Carbohydrate Chemistry* (Alexander, R.J., ed.) American Association of Cereal Chemists, St Paul, Minessota, p. 37.

Hough, M.C. & Wright, D.C. (1996) *Polym. Test.* **15**, 407.

Johannaber, F. (1985) *Injection-Moulding Machines*, second edition, Hanser, New York, p. 39.

Jordan, N.D., Bassett, D.C., Olley, R.H. & Smith, N. (2000) *J. Biomedical Mater. Res.*, in press.

Kalay, G., Alan, P.S. & Bevis, M.J. (1997) *Kunststoffe* **87**, 768.

Kalay, G., Sousa, R.A., Reis, R.L., Cunha, A.M. & Bevis, M.J. (1999) *J. Appl. Polym. Sci.* **73**, 2473.

Kucner, L.K. & McManus, H.L. (1994) *Int. SAMPE Tech. Conf.* **26**, 341.

Lee, E.H., Rado, G.R., Mansur, L.K. & Zachmann, H.G. (1996) *Trends Polym. Sci.* **4**, 229.

Lin Jana, J. & Shen, J.J. (1993) *Carbohydrate Res.* **247**, 279.

López Cabarcos, E., Zachmann, H.G., Bayer, R.K., Baltá Calleja, F.J. & Meins, W. (1989) *Polym. Eng. Sci.* **29**, 1983.

Morcher, E. (1984) in *The Cementless Fixation of Hip Endoprostheses*, Springer Verlag, New York, p. 1.

Nachtmann, Ch. (1996) *Galvanotechnik* **87**, 3702.

Nakashima, T. & Matsuo, M. (1996) *J. Macromol. Sci., Phys.* **B35**, 659.

Olley, R.H., Hosier, I.L, Bassett, D.C. & Smith, N.G. (1999) *Biomaterials* **20**, 2037.

Paar, A. (1988) *UMHT-3 Ultra Microhardness Tester. Polypropylene Pipes* **107**, 5.

Paar, A. (1990) *UMHT-4 Ultra Microhardness Tester. Individualization of Human Hair* **111N**, 11.

Paplham, W.P., Seferis, J.C., Baltá Calleja, F.J. & Zachmann, H.G. (1995) *Polymer Composites* **16**, 424.

Parvatareddy, H., Wang, J.Z., Dillard, D.A. & Reifsnider, K.L. (1996) *J. Compos. Mater.* **30**, 210.

Pivin, J.C. (1994) *Nucl. Instrum. Methods Phys. Res.* **B84**, 484.

Poutanen, J. & Forssell, P. (1996) *Trends Polym. Sci.* **4**, 128.

Rao, G.R., Blau, P.J. & Lee, E.H. (1995) *Wear* **184**, 213.

Rao, G.R., Lee, E.H. & Mansur, L.K. (1992) *Mater. Res. Soc. Symp. Proc.* **239**, 189.

Rao, G.R., Monar, K., Lee, E.H. & Teglio, J.R. (1994) *Surf. Coat. Technol.* **64**, 69.

Rose, P.I. (1987) in *Encyclopedia of Polymer Science and Engineering* 2nd edition, Vol. 7 (Mark, H.F., Bikales, N.M., Overberger, C.G. & Menges, G. eds.) John Wiley & Sons, New York, p. 488.

Rueda, D.R., Ania, F. & Baltá Calleja, F.J. (1997) *Polymer* **38**, 2027.

Rueda, D.R., Baltá Calleja, F.J., Ayres de Campos, J.M. & Cagiao, M.E. (1988) *J. Mater. Sci.* **23**, 4487.

Rueda, D.R., Baltá Calleja, F.J. & Bayer, R.K. (1981) *J. Mater. Sci.* **16**, 3371.

Rueda, D.R., Baltá Calleja, F.J., García, J., Ward, I.M. & Richardson, A. (1984) *J. Mater. Sci.* **19**, 2615.

Rueda, D.R., Bayer, R.K., Baltá Calleja, F.J. & Zachmann, H.G. (1989) *J. Macromol. Sci. Phys.* **B28**, 267.

Secall, T. (1996) Master's Thesis, University of Kassel, Germany.

Shogren, R.L. (1992) *Carbohydr. Polym.* **19**, 83.

Turnbull, A. & White, D. (1996) *J. Mater. Sci.* **31**, 4189.

Vassileva, E., Baltá Calleja, F.J, Cagiao, M.E. & Fakirov, S. (1998) *Macromol. Rapid. Common.* **19**, 451.

Yannas, I.V. & Tobolsky, A.V. (1967) *Nature* **215**, 509.

Waigh, T.A., Hopkinson, I.M., Donald, A.M., Butler, M.F., Heidelbach, F. & Riekel, C. (1997) *Macromolecules* **30**, 3813.

Ward, A. & Courts, A. (eds.) (1977) *The Science and Technology of Gelatin*, Academic Press, London.

Wilkinson, A.N. & Ryan, A.J. (1999) *Polymer Processing and Structure Development*, Kluwer, Dordrecht.

Zeleznak, K.J. & Hoseney, R.C. (1987) *Am. Assoc. Cereal Chem.* **64**, 121.

Author index

Subject index